Biomedical Research

How to plan, publish and present it

Springer
London
Berlin
Heidelberg
New York
Barcelona
Budapest
Hong Kong
Milan
Paris
Santa Clara
Singapore
Tokyo

BIOMEDICAL RESEARCH

How to plan, publish and present it

William F. Whimster
with contributions from Gary Horrocks and
David A. Heath

Foreword by Professor Alex Bearn

William F. Whimster, MA, MD (Cantab), FRCP (Lond), FRCPath
Professor of Histopathology, The School of Medicine and Dentistry, King's College, London.

Co-author
Gary Horrocks, BA, MSc, Assistant Librarian (Information Services), The School of Medicine and Dentistry, King's College, London.

ISBN 3-540-19876-8 Springer-Verlag Berlin Heidelberg New York

British Library Cataloguing in Publication Data
Whimster, W. F. (William Frederick)
Biomedical research : how to plan, publish and present it
1.Biology – Research 2.Medicine – Research
I.Title
610.7'2041
ISBN 3540198768

Library of Congress Cataloging-in-Publication data
Whimster, W. F. (William Frederick)
Biomedical research : how to plan, publish, and present it / William F. Whimster.
p. cm.
Includes bibliographical references and index.
ISBN 3-540-19876-8 (paperback : alk. paper)
1. Medicine--Research. 2. Medical sciences--Research. I. Title.
R850.W47 1996
610'.72--dc20 96-42694

Set in 10/11pt Times by Editburo, Lewes, East Sussex
Printed and bound at Interprint Ltd., Malta
28/3830-543210 Printed on acid-free paper

Contents

Acknowledgements

Springer gratefully acknowledges the use of the following:

Figure	Credit
Fig. 1.1	Reproduced by kind permission of the photographer – Ross Bird
Table 4.2	Reproduced by kind permission of BMJ Publishing Group 1995
Table 4.4	Reproduced by kind permission of the *Journal of the National Cancer Institute*
Fig. 4.6	Reproduced by kind permission of Springer-Verlag Heidelberg Berlin, *Diabetologia*
Fig. 4.7	Reproduced by kind permission of Springer-Verlag Heidelberg Berlin 1991
Fig. 5.1	Reproduced by courtesy of the Trustees, The National Gallery, London
Fig. 5.11	Copyright, 1942, *Los Angeles Times*. Reprinted by permission
Fig. 5.13	Reproduced by kind permission of Butterworth-Heinemann 1994
Fig. 8.4	Reproduced by kind permission of Peanuts – © 1981 United Features Syndicate, Inc
Fig. 12.1	Reproduced by kind permission of Nuffield Provincial Hospitals Trust 1985
Fig. 12.2	Reproduced by kind permission of Nuffield Provincial Hospitals Trust 1985

Foreword

It is a distinct pleasure to be invited to prepare a short Foreword to *Biomedical Research: How to plan, publish and present it*, by William F. Whimster. Ninety years have elapsed since T. Clifford Allbutt, the Regius Professor of Physic at the University of Cambridge, published his classic work of 1904 *Notes on the Composition of Scientific Papers*. Small in size, but deep in wisdom, it remains a remarkably useful, if slightly old-fashioned, book, still well worth reading.

Since 1904, and particularly in the last 25 years, there has been an avalanche of books on scientific style. Medawar has aptly observed that *"most scientists do not know how to write, insofar as style betrays l'homme même, they write as if they hated writing and wanted nothing more than to have done with it."* Whimster's book has a broader objective than most of this genre. Unlike Allbutt, who was addressing in the main those who were writing their theses to obtain the MD, Whimster writes for the young medical scientists who are planning and writing up an account of their research, either for publication in scientific journals, or for presentation of the scientific material at meetings. Whimster, a scientist and an experienced long-term science editor, has written an up-to-date version of an earlier and very successful volume, *Research, How to Plan, Speak and Write About It*, edited by C. Hawkins and M. Sorgi. One of the principal virtues of the earlier volume and now superbly maintained and expanded by Whimster is its clarity and common sense.

For the novice, examination of the early chapters will provide a vital introduction on how to formulate and approach a scientific problem. Matters of syntax and style are presented without pomposity and even the initiated can read them with profit and pleasure. Nevertheless, Whimster makes it abundantly clear that the approach to a scientific problem is of cardinal importance. Even the most sophisticated of biometricians cannot salvage a poorly designed experiment, as Whimster properly emphasises. The short introductory section on statistics is both succinct and admirably clear.

In the last analysis, however, it is hard to avoid the conclusion that the most reliable and the most enjoyable way to learn how to design an experiment and how to write up the results is to be apprenticed to a mature senior scientist in an active, well recognised laboratory. The scientific and literary discipline imposed by a senior scientist, at an early stage of a scientific career, is of unparalleled value. One only has to read the papers of some of the masters – Lord Adrian, J. B. S. Haldane, Thomas Lewis, Douglas Black and Peter Medawar, to realise that a scientific experiment can be both beautifully designed and unambiguously and clearly presented. Thus, particularly for those less fortunate in their scientific mentor, this book will undoubtedly lead to more thoughtful and well planned research as well as a clearer presentation of the results.

This carefully written book will improve the design and presentation of the results of scientific research. It can be recommended enthusiastically and without reservation. It will not, of course, provide any guarantee of scientific success, but it contains a wealth of wise, unpretentious advice, deftly and agreeably written. Even the accomplished scientist will gain from leafing through this book, and there is no doubt that scientific presentations, both written and oral, will be considerably clearer if the gentle admonitions of Whimster are remembered and heeded.

Alexander G. Bearn, MD, FRCP, FRCPEd, FACP
Professor Emeritus of Medicine,
Cornell University Medical College, Ithaca, New York, USA;
Formerly Senior Vice President, Medical and Scientific Affairs,
Merck Sharp and Dohme International, Rahwey, New Jersey, USA.

Preface

Clifford Hawkins and Marco Sorgi believed that those doing research needed a book which dealt "with all the steps that are necessary from the start to the finish", apart from the initial idea and doing the actual research. The success of their 1985 publication, Research: how to plan, speak and write about it, confirmed their belief. Recent market research also confirms that those tackling BSc, MPhil, MSc and PhD projects and theses, setting off on the road to their first publications in the biomedical "literature", or writing grant applications, would welcome an updated book. As so much has moved on I have largely rewritten the book rather than re-editing it. It is interesting, however, that David Heath's chapter "Research: Why do it?" remains as pertinent today as it was in 1985. He managed to touch on all the issues facing research in biomedicine that continue to exercise the profession, governments and society throughout the world, with some agencies doing something about some of them and making some worse, as my additional notes show.

There have been many developments in the areas covered by the 1985 book. Many, if not most, of today's readers are now familiar with the use of computerised word processing, graphics and statistics packages. Library usage has been transformed by new computerised facilities, both within the libraries and on-line to departmental and home computers through the various networks and databases. The 1978 Vancouver Group of editors, now the International Committee of Medical Journal Editors, has harmonised and given worldwide authority to "Instructions to Authors", as well as to international policies on many contentious publishing issues (Appendix 1). Individual journal editors have added their own points, and the British Medical Journal in particular has worked on guidelines for referees, statistical advisers and technical editors (Appendix 2), which authors can use to see the view from the journals' side. Indeed, authors who neglect these two appendices risk rejection just as surely as if they had done bad research. Much else has also been written since 1985: references specifically applicable to points in the text are listed at the end of the relevant chapter. Other useful material is listed in Further Reading at the end of the book.

The practice of biomedical research has become fairly standardised throughout the world, with, for example, worldwide declarations such as the Declaration of Helsinki to regulate the conduct of research on human beings (Appendix 3). Other regulations, for example in relation to animal experimentation, health and safety, access to information resources, grant applications and so on, do of course differ from country to country. Nevertheless, the principles now established are widely understood and applicable. Although UK practices are used as examples in this book, readers in other parts of the world, such as the USA, Asia and the Far East, should have no difficulty adjusting what is said here to dealing with their own authorities or indeed establishing easy communication with the international biomedical community.

For directness and simplicity I wish through these pages to speak to individual doctors/research workers/readers/writers, whom I have addressed as "you", and tried to imagine the dialogue between us. I also appreciate that "you" are just as likely to be a man or a woman, but have used "he" and "his" throughout to stand also for "she" and "hers" because the alternatives make for such clumsy writing and reading. I have assumed that you work in the related and overlapping fields of biological or medical science, and have therefore used the rather inelegant but useful term "biomedical".

"Publication" includes presenting your work at workshops, meetings, seminars, symposia, conferences or congresses, in ascending order of stage-fright-inducing grandeur. Success in these circumstances also depends on planning, together with knowing how to get on the programme and how to present both spoken papers and posters. Advice is available in later chapters.

It is sad that Clifford Hawkins died in 1991 (obituary BMJ 4 January 1992;304:51). He did enjoy all the new things. His friendship, fertile imagination, enthusiasm and splendid laugh are greatly missed. This new volume builds on his work. Nevertheless, although I have drawn on his ideas and those in the many other books on medical writing, what I say is what I believe from my own experience of writing and presenting in my own fields. Many points may be disputed or refuted by the experience of others in mine or other fields. My aim has been to include not only planning for writing and research (Hawkins's idea) but also the reasons for choosing one way of doing it rather than another. This is in distinction to those books which give examples of writing which are designated as "good" or "bad", "interesting" or "boring", without saying in what way these epithets are applicable. Such books without rationale encourage readers to follow the examples slavishly or intermittently (depending on their memories) but do not help them to create their own options and styles.

Finally I wish to express my thanks for the help I have had from those who contributed to the previous edition; to Stephen Lock, Yrjo Collan, Alex Paton, Tim Albert, my friends at the BMJ and many other colleagues and course participants for sharing their ideas over many years; to Professor Norman Noah, for reading the epidemiology and statistics; Chris Hogg, Gary Horrocks and the library staff at King's College School of Medicine; to Barry Pike, Alex Dionysiou and the King's Healthcare Photographic Department staff and Ray Thatcher in the Comparative Biology Unit; and to Dr Gerald Graham, without whom I would never have got around to moving a word-processing finger, as well as the staff at Springer-Verlag.

William Whimster

Introduction: Research – why do it?

David A. Heath FRCP, Reader in Medicine,
University of Birmingham
(with additional notes by William Whimster)

The faculties developed by doing research are those most needed in diagnosis.

FH Adler (1966)

Many are the motives for doing research. Some are tempted by the excitement of discovery – the chance of adding something new, however small, to the expanding frontier of knowledge. This may be an inherent feature of the human intellect and may explain much of the progress along the laborious road of biological evolution. A few have this quality in excess and are self-propelled; they are undeterred by drudgery and disappointment, however great these may be. Others are stimulated by more mundane though important objectives: the pursuit of prestige, or the need for publication due to the publish-or-perish pressure when climbing the ladder to a successful career.

Whatever the reason for undertaking research, the benefits are undoubted:

- A critical or scientific attitude is developed.
- The chance to study a subject in depth.
- Getting to know how to use a library.
- Learning to assess the medical literature critically.
- Development of special interests and skills.
- Understanding the attitude of others, whether in routine or research laboratories.
- Obtaining a higher degree.

Discovery – by forethought and serendipidity

New knowledge arises in various ways. It may originate in some quite unexpected observation which occurs during an ongoing investigation in an academic department – providing this is fully exploited. Opportunities come more often to active bench workers and to those involved in new techniques or with new equipment, for example an electron microscope.

Serendipidity is the faculty of making happy discoveries by accident and is derived from the title of the fairy tale The Three Princes of Serendip, the heroes of which were always making such discoveries. The role of chance in research has been discussed by several authors and many examples of discoveries where chance played a part are given by Beveridge (1950) in The Art of Scientific Investigation. James Austin (1978) in Chase, Chance and Creativity analysed the varieties of chance that contribute to creative events:

Chance i This is the type of blind luck which can provide an opportunity to anyone motivated to do research. An example is the unexpected arrival of a patient with a rare metabolic disorder.

Chance ii Here something has been added: action. Austin quotes Kettering, the automotive engineer, who stated "keep on going and the chances are that you will stumble on something, perhaps when you are least expecting it. I have never heard of anyone stumbling on something sitting down".

Paul Ehrlich provides a good example of this. He fervently believed in the possibility of a chemical cure for syphilis, persisting when all reasonable hope had gone – and finally succeeded with Salvarsan, the 606th compound to be tested. More recent was the discovery, for the first time, of a virus causing malignant disease in humans: Epstein and Barr persisted in spite of many failures in searching for this in the tissues of Burkitt's lymphoma, eventually culturing the cells and seeing the virus under the electron microscope.

Chance iii Chance presents only a faint clue; the potential opportunity exists, but it will be overlooked except by that one person uniquely equipped to grasp its significance. As Louis Pasteur immortalised it, "Chance favours only the prepared mind". For example, although Sir Alexander Fleming discovered penicillin by

Fig. I.1. William Harvey.

serendipity in 1929, it was not until 1939 that Florey and Chain, working at Oxford, revealed its practical importance.

Chance iv Here there is "one quixotic rider cantering in on his own homemade hobby horse to intercept the problem at an odd angle". This type of discovery is usually due to a combination of persistence and lateral thinking.

Historical Aspects

Every research worker should have some knowledge of the history of original investigation. Space does not allow more than mention of the landmarks, and many articles have been written about the remarkable information explosion due to research in this century (Weatherall, 1981). Initially most medical research was started by individuals who spent many years either experimenting or collecting clinical information. One of the first to demonstrate the value of experimentation in clinical medicine was William Harvey (fig. I.1), who discovered the circulation of the blood in the seventeenth century. As with many of the great medical pioneers, he was ridiculed at the time and it was many years before the funda-

Fig. I.2. Thomas Sydenham.

mental discoveries that he made affected medical thinking and practice.

The tradition of clinical observation at the bedside was far more popular than experimentation, and this has been the mainstay of medical knowledge until the present century. One of the early proponents of this approach was Thomas Sydenham (fig. I.2), who has been called the 'English Hippocrates'. He started medicine late, and qualified at the age of 39 after being a captain in Oliver Cromwell's army. He is justly famous for his studies of malarial fevers, dysentery, scarlet fever, measles and the chorea which bears his name. His best known work is his treatise on gout, from which he suffered. His account of hysteria, which he claimed affected half of his non-fever patients – today called psychosomatic disease – is a masterpiece of sober description. His treatment was relatively reasonable and he based it on supporting the vis medicatrix naturae (the healing power of nature), though he did not escape entirely the temptation to do extensive blood-letting. He also deserves credit because, although he was a Puritan, he did adopt the new wonder drug, quinine, the "Jesuit powder", imported from Peru in the 1630s; apart from curing malaria (the most frequent disease of the time), this allowed it to be separated from other fevers.

Fig. I.3. William Withering.

The eighteenth century brought with it some of the first major therapeutic discoveries: William Withering (fig. I.3) of Birmingham, who was a clinician, botanist and social reformer, introduced digitalis into orthodox medicine after learning of the use of the plant foxglove for dropsy from an old woman in 1779; this clinical trial greatly advanced the treatment of patients with heart failure. A cure was found for scurvy, and Jenner began his great work on vaccination against smallpox; he was a country doctor and had heard of the immunity against smallpox enjoyed by milkmaids who had previously been infected by cowpox. With the encouragement of his teacher, John Hunter, he started a research project on this and published his article in 1789, in which he demonstrated that inoculation with cowpox would produce protection against smallpox in man without ill effects to the patient. In spite of the fact that the paper was rejected by the Royal Society (Chapter 15), Jenner's discovery went on to be of incalculable benefit to mankind, and the World Health Organisation has now found it possible to announce the "eradication" of smallpox. Careful scientific work was beginning genuinely to benefit man.

The experimental approach of the eighteenth century waned in

Fig. I.4. Claude Bernard.

Fig. I.5. Robert Koch.

England in the nineteenth century and failed to sustain the impetus of the previous century, with most major contributions being descriptive ones (Booth, 1979). However, in Europe, medical research was gaining momentum, particularly as microscopy developed. Claude Bernard's (fig. I.4) research greatly advanced our knowledge of human physiology, and Pasteur performed experiments which indicated the existence of microorganisms as the cause of fermentation. Robert Koch (fig. I.5) who, unlike Pasteur, was medically qualified, subsequently proved that an organism could cause a specific disease. He used solid media and developed new methods of fixing and staining bacteria. Unfortunately, a wave of uncritical research then started, so he issued his famous postulates as criteria for further valid research: (1) The organism should be found in each case of the disease. (2) It should not be found in other diseases. (3) It should be isolated. (4) It should be cultured. (5) It should, when inoculated, produce the same disease. (6) It should be recovered from the inoculated animal. However, as in original work today, the ideal cannot always be achieved.

Research in the twentieth century

The present century began with medicine being predominantly a descriptive art with few effective remedies. Much was to change

Fig. I.6. Sir Thomas Lewis.

over the next 80 years, a great deal of which can be traced back to the development at the beginning of the century of specific medical research institutes which encouraged and financed full-time medical research. In the United States of America the Rockerfeller Institute was created in 1901. In England, Sir Thomas Lewis (fig. I.6), working for the Medical Research Council, encouraged the development of clinical research as a career. The first biography of Sir Thomas Lewis, written by Arthur Hollman, has just been published (Hollman, 1996). Flexner (1912) commented on the educational systems in various countries and noted that most research appeared to be carried out in places where the staff worked full time in the hospital or institution. This important observation was noted by the Haldane Commission (1910–1913) and must have greatly influenced its members, for they then recommended the creation of full-time professors to direct academic or research departments. The greatest impetus for the rapid growth of medical science was the commitment by governments of major sums of public money to scientific medical research. The best example of this was the progressive development of the National Institutes of Health in America, which initiated research into widely ranging areas of medicine and the basic sciences.

The information explosion

Fredrickson (1981) has outlined the enormous developments of the past 30 years, when an exponential increase in our knowledge of the basic sciences has been matched by major advances in the treatment of patients, and has enabled medicine to develop truly into both a descriptive and a therapeutic speciality. The very nature of these developments has brought with it tremendous problems. The range of most medical research is now so great as to be beyond the comprehension of the single person. Its cost has increased dramatically, leading to considerable problems in funding. More recently, the potential commercial implications of university-based research have raised interesting new problems concerning the relationship between research institutes and profit-making organisations.

Financial problems

Today we find scientific research threatened by the decreased affluence of the industrial nations. Research which, in many areas, has become increasingly expensive is having to share the pruning that is going on in most walks f life. Traditional sources of funds are being reduced. The various types of research and how these should be funded need to be discussed. Each may require a different type of research worker and different forms of research may be more appropriate for different countries, depending on their affluence and disease patterns.

Note: However, more money is being put into biomedical research. In 1995 the government of the UK formally allocated 1.5% of Gross National Product, i.e. tax payers' resources (about £600 million annually) to NHS research and development. The House of Lords' Select Committee on Science and Technology identified four perceived categories of research:

- *suitable for central funding*
- *not suitable for central funding but of sufficient value or potential value to the NHS in the short or long term to warrant continuing support*
- *pre-protocol and curiosity-driven*
- *not worth supporting.*

It noted that two other kinds of research might suffer if service support were conditional on benefit to the NHS, namely, research which might lead to more expensive health care, and research in areas of more interest overseas than in the UK. In the private sector the giant Wellcome organisation liquidated many of its assets in order to put more resources into research. Nevertheless, demands for research funds are ever increasing, and so is the bureaucracy surrounding their allocation (Chapter 5). It will never be possible for anyone, newcomers to research or old hands with good fundraising records, to know whence, if at all, the next grant will come from. That is part of a research worker's life and stimulation to further efforts.

Types of research

Weatherall (1981) distinguishes between basic research, applied research and development, and clinical trials and the monitoring of the fruits of research in everyday use.

Basic research

Basic research has been fundamental to most of the major medical advances ever made. It differs from all other forms of research in being totally unpredictable, and often there is no initial connection between the research and its medical application. Indeed, the most important scientific discoveries have been made by investigators pursuing their own ideas (Fredrickson, 1982). Medical application may follow, but it may take a long time before this happens. Harvey's discovery of the circulation of the blood had no discernible effect upon medical practice for almost 300 years (Bearn, 1981). By its very nature, basic research requires a major commitment from researchers who must have a good training in research techniques and in fields outside medicine. Results are unlikely to be achieved rapidly, so that a long-term commitment by those involved as well as by funding bodies is essential.

Unfortunately, the number of medically qualified people involved in this work appears to be falling, as in other areas of medical research. Physician post-doctoral researchers in the US have declined from 65% of the total National Institute of Health trainees and fellows in 1970 to 30% in 1980 (Fredrickson, 1981), and a similar change is probably taking place in Britain (Peart, 1981). This is partly due to the increased complexity of medical research, though perhaps more importantly to the uncertainties of a career in research compared with medical practice and its greater remunerations; added to this is the increasing rigidity of undergraduate and postgraduate medical education, which makes it far more difficult to divert a significant amount of time to organising research. The enormous financial cost and complexity of basic research will probably demand that it becomes more and more localised to fewer big institutions, especially universities. Until possible practical applications emerge, the work is less likely to be supported by industry or private sources.

Applied and developmental research

Applied and developmental medical research must include a major contingent of medically qualified people to link the scientific advances to patient care. For this to be readily achieved, there will need to be close links, ideally physical, between the centres developing the new process and the patients; the linkage between universities and clinical medical schools is a good example of such a union. Wherever possible, the nucleus of the development team would be medically qualified, but also scientifically trained, people. There is, however, a much greater possibility for doctors to spend a fruitful shorter period in research before returning to clinical practice. Such a scheme should ensure that the worker joined an established work team and thus was able to be educated in research techniques while following a proven field of interest.

Clinical trials and monitoring of research consequences

Clinical trials and monitoring the fruits of research in everyday clinical practice represent a different type of research work. By the time new developments have reached the clinical level, commercial organisations have usually become involved. This has various ramifications: first, the commercial organisation has a major commitment to having its product used, often even before its value has been demonstrated unequivocally; second, such an organisation often funds researchers to try to prove the value of the product, though all too frequently it fails to justify the original hopes when good evidence is obtained.

Audit on research

Not surprisingly, there has been little "research on research". Comroe and Dripps, however, studied the research that led to ten major clinical cardiovascular and pulmonary advances between the period of 1945 and 1975 (Ringler, 1977). Having decided on the key articles for these, they noted that over 40% of the research had a goal unrelated to the later clinical advance. Over 60% of the research was of a basic nature and nearly 70% had been performed in colleges, universities and medical schools. In almost 60% of cases it took over 20 years for the original research to lead to the clinical advance. This study highlights the need to continue to support basic research, which usually will not be specifically directed at clinical problems. The link of basic research to clinical institutes is also stressed as being essential for the appropriate application of advances to patient care. The information produced by this report also suggests that further "research on research" may be fruitful.

Note: In the 1990s the lack of "research on research" has been addressed in many countries whose authorities wish and need to know what research is going on. In the UK the Higher Education Funding Council has instituted "research assessment" exercises to do this, on the basis of which all university departments, including medical school departments, are graded on a 1–5 scale and will receive funding for research accordingly. Many of us believe that the methodology for the assessments is very crude and unreliable, not least because it is based more on quantity of output than "quality", whatever that is in research.

Who should do research?

The question is whether all doctors or merely a selected few should perform research. Obviously not all doctors can be involved in basic research; this occupies virtually all of the person's time for a considerable duration, and the fact that much of the work will be far removed from clinical practice may make it an unsuitable field for those wishing to spend a short time in full-time research. Furthermore, the need for the research worker to be appointed for prolonged periods in itself poses problems, for the most innovative and active phase of a research worker's life is likely to be during his or her younger years – "if you block the tide of the young coming in, you destroy the vitality of science" (Fredrickson, 1982). To date we have not been able to devise a scheme for dealing with the research worker who is no longer as productive, though not yet ready to retire. As stated earlier, it is likely that more and more basic research will be carried out in fewer, larger institutes. Funding will be predominantly from governmental resources, with industry playing a minor role. These full-time investigators ideally will be left to pursue their own ideas, with often no obvious benefit to patients in mind. Some medically qualified people will enter such areas, but they will probably retain little if any contact with patients.

The vast majority of medically qualified doctors will want to spend a major part of their time in active clinical practice and only devote two or three years to part-time or full-time research, preferably in a laboratory.

Note: There is a small percentage of young doctors who naturally undertake research. They will always find ways of doing it. Some find they like research after being pushed into doing some for their career progress, and continue to do it. The remainder never really get into it but carry on happily being doctors. The postgraduate training of specialists in the UK has been overhauled recently, and generally shortened, with more organised programmes and supervision. I expect this to further reduce the inclination of the third group of young doctors to do research seriously. This needs flexibility, time and freedom, not more bureaucracy, the work of which falls on other doctors and impedes their research too.

Should all doctors do some research?

There is a strong argument that all doctors should do some research, even if they wish to spend the rest of their lives dealing with patients. A scientific approach is essential, for, as medicine advances rapidly, so the need for a more critical evaluation of new developments becomes more urgent. The more potent the treatments or new procedures we develop, the greater the chances of benefit and harm to the patient. The medical past is littered with example after example of possible major advances eventually being shown to be of no value or, much worse, of positive harm to the patient. Despite these warnings, the medical profession and the general public seem to wish to try more and more unproven treatment. The problems of the past are likely to keep repeating themselves unless a far greater critical attitude develops within our profession. Hence, for medical advances to continue, research will always need to be done, and for the high standards of medicine to be applied, a carefully developed critical mind is needed. This is most likely to be developed in a research environment. Here the doctor will be constantly exposed to the critical evaluation of previous work, the design of good projects and the discipline of writing up observations. Also the opportunity to review papers under supervision often arises; this greatly adds to the ability to evaluate new claims more carefully, as do the rigours of presenting work to a critical audience.

References

Austin JH. Chase, chance and creativity: The lucky art of novelty. New York: Columbia University Press, 1978.

Bearn AG. The pharmaceutical industry and academe: Partners in progress. Am J Med 1981;71:81–8.

Beveridge WB. The art of scientific investigation. London: Heinemann, 1950.

Booth CC. The development of clinical science in Britain. Br Med J 1979; 1:1469–73.

Flexner A. Medical education in Europe. A report to the Carnegie Foundation for the advancement of teaching. Bulletin No. 6, 1912.

Fredrickson DS. Biomedical research in the 1980s. N Engl J Med 1981; 304:509–17.

Fredrickson DS. 'Venice' is not sinking (the water is rising). Some views on biomedical research. JAMA 1982; 247:3072–5.

Haldane JS (Chairman). Report to the Royal Commission on University Education in London (1910–1913).

Hollman A. Sir Thomas Lewis: Pioneer Cardiologist and Clinical Scientist. London: Springer, 1996.

Peart WS. Advice from a not so young medical scientist. Clin Sci 1981;61:364–8.

Ringler RL. The Comroe–Dripps report on the scientific basis for the support of biomedical science. Fed Proc 1977;36:2564–5.

Weatherall M. Medical research and national economics. J R Soc Med 1981; 74:407–8.

Reprinted from *Research, How to Plan, Speak and Write About It* by kind permission of David Heath, with additional notes in italic by William Whimster.

A
PLANNING

1. Types and design of research

The probability that a paper with a clear image will emerge from research is determined more by how the research was conceived and planned than by how well the paper is written.

Ed Huth, Editor, Annals of Internal Medicine

Any human activity that consists of discovering or deciding that you want to know something, turning what you want to know into a question, somehow answering the question and then conveying the answer to others, is research. Why you might want to pursue such an activity is a separate question that is discussed below.

For our purposes, the activity may be reformulated as: having a research idea, formulating the question to be answered or the hypothesis to be tested, doing the work needed to obtain the answer, publishing and presenting the results of the work.

The research idea

This book cannot help you to decide on an idea for research. Most biomedical people are aware of what they would like to know in their areas of interest, and have plenty of ideas. Some, however, do not know whether what they would like to know is already known. These people have to start by finding that out, through literature searches or in some other way. Others do not know whether it is actually possible, with current knowledge and techniques, let alone with what may be available to them personally, to find out what they want to know; this needs careful thought, since other people, such as funding or ethics committee members, may decide for you that it is not possible.

Other readers may be looking for a research idea. They can read the papers in the journals of their areas of interest, perhaps concentrating on the discussions and introductions, in which authors often point out where further research is needed, or indeed spotting errors or omissions in the methods or results, which may indicate work that could usefully be repeated or extended. Textbook authors usually give the impression that there are no gaps in knowledge, so it is difficult, but not at all impossible, to get research ideas from them. Another alternative is to ask any available research or service worker in the field what knowledge they are lacking, or whether they have a project waiting to be done.

Formulating the question or hypothesis

As an example, the general idea may be that it would be interesting/useful/essential to know whether tumour blood vessel proliferation (angiogenesis) is an important factor in the prediction of tumour growth and metastasis in non-small cell lung cancer, because, if it is, it may be possible to suppress the angiogenesis and prevent growth and/or metastasis. This is too complicated an idea for one piece of research.

The idea could be reformulated into a first question: "Is the amount of tumour angiogenesis, measured in some way, correlated with tumour growth and metastasis in non-small cell lung cancer?" The possible outcomes of the work are then "yes", "no", or yes or no with qualifications, or "don't know", which still leaves the reader floundering somewhat.

On the other hand, the hypothesis form would be: "*Because* angiogenesis is essential for tumour growth and metastasis formation (or any other reasons you may find), my hypothesis is that I shall find more angiogenesis in rapidly growing and/or metastasising non-small cell lung cancers." The outcomes would then be more clear cut: I did or did not find more angiogenesis in rapidly growing and/or metastasising non-small cell lung cancers, and furthermore the angiogenesis did/did not correlate with the rate of tumour growth and/or metastasis. Thus the hypothesis formulation, with the built in "*because*" logic, makes the work to be done clearer for the researcher, and he will process it into print more clearly for the reader as a result.

It also becomes clear what type of research you are heading for. If it appears that similar research appears to have been done before, you can bring into your hypothesis why you should do more cases or additional investigations; but the results are never exactly the same, and further cases from a different centre strengthen or weaken the hypothesis until the truth of it becomes first convincing and then irrefutable.

Types of research

Research has been classified as "basic" or "applied". In the biomedical field research consists of making observations on some form of biomedical material, performing clinical trials on patients, or carrying out experiments on animals or *in vitro*.

Observational research

Darwin's research was observational. He observed, collected and dissected many individual animals, particularly during the voyage of the *Beagle*, including finches, at least one of Darwin's being still extant (fig. 1.1; Hawkes, 1966). In the same way, recent work by the Grants on finches in the Galapagos islands is observational (Weiner, 1994) in that defined groups of finches were observed in relation to specified characteristics over specified times in specified circumstances, and conclusions were drawn without interfering with any of the characteristics or circumstances under observation.

Fig. 1.1. One of Darwin's finches.

Case reports are often the result of observations made on one patient and may have far reaching implications, for example the understanding of the disease caused by the deficiency of clotting factor IX (Christmas disease) from the observation of a patient called Christmas. One observation was enough to lead Fleming to the discovery of penicillin. To say that "chance favours only the prepared mind" (p. xiv) means being alert to the unusualness of any observation, such as Sherlock Holmes noticing not only that the dog did not bark in the night but that that observation was of significance, albeit clearly not of statistical significance.

Obsessional recording of observations both helps and hinders the process. On the one hand the observations can be retrieved accurately, on the other a significant one may be obscured by all the insignificant ones.

Many a newcomer's first publication in the biomedical field has been a case report. Unfortunately case reports are commonly rejected, either because the observations or combinations of observations reported are not original, that is, they have been reported before, or because no clinical point is made, the referee applying the "So what?" test. Just to make observations is therefore not enough; the case must be searched for its originality and for what enlightenment the observations might bring to physiology or pathology, or for how they might enhance clinical practice or understanding; or if none of these, further observations that might do so can be sought, sometimes in vain.

Further research often involves the study of groups of cases to see how often, when and where the initial individual observations or diseases occur, so that one can determine why they occur. This is then epidemiology. Epidemiologists have developed rigorous techniques for the collection and analysis of their data, but their work is still mainly observational, although intervention studies, for example the effects of changing diet, are performed. Their techniques are used not only to study patients but also to observe any specimens, for example lung weights at postmortem, or images, for example the growth of a tumour in successive chest x-rays, or cells, for example the number that exhibit proliferation markers on staining.

Epidemiological surveys generally identify associations without being able to prove that they are causal associations. For example, an increase in ownership of television sets in a population may be associated with an increase in the incidence of coronary artery disease, but this observed association does not prove that ownership of a television set causes coronary artery disease. A causal association becomes more likely and convincing if successive surveys in different populations all show the same association, as was the case with the association between cigarette smoking and lung cancer. Bradford Hill (1991) sets out the criteria for causal association.

There are several types of epidemiological survey. In *longitudinal studies* variables such as risk factors or health outcomes are observed over time in groups or "cohorts" of subjects. Large and lengthy studies may be needed to investigate low-incidence events or associations. A quicker alternative may be the *case-control study* in which patients with the disease under study are compared with controls who do not have it. Case selection and allowances for potential confounding factors are obviously very important. *Cross-sectional* studies measure the prevalence of variables, such as television set ownership and coronary artery disease, in a population at a specific time.

It is encouraging that solitary individuals with little equipment can still make observations on their own that lead to leaps of understanding, as Burkitt showed (1958) in his studies of his eponymous lymphoma in Africa. This is often in the face of scepticism, as shown by the work of Wyllie and his colleagues (Kerr *et al.*, 1972) in their microscopical studies of apoptosis, "programmed cell death".

In spite of these and many other solo observational achievements, observational research comes lowest in the hierarchy of biomedical research prestige. It has nevertheless been the principle biomedical research method of the naturalists and physicians of previous centuries, and a large proportion of our knowledge comes from it even today, particularly in epidemiological research and in the laboratory and imaging departments. As my father used to say, "many more misunderstandings come from not observing than from not knowing"; organised and focused observing is still a valuable research tool. Although few newcomers to research are likely to start off with an epoch-making original observation, the original observations usually lead to further observational research which newcomers or their supervisors may appreciate the need for; this has to be carefully planned, not least because others in your "invisible college" (Chapter 17), are likely to have the same research idea. On the other hand, two pieces of observational research in the biomedical field seldom come up with exactly the same answers; they may support each other or they may be contradictory. Either way, they both contribute to the body of knowledge, and both are publishable. The *idea* may not be original when the *results* are.

Clinical trials

Put on to a firm statistical and clinical footing by such eminent enthusiasts as medical statistician Sir Austin Bradford Hill (1937, latest edition 1991) and physician Archie Cochrane (1972, latest edition 1989), after whom the recently established Cochrane Centre for "evidence based medicine" in Oxford is named, clinical trials carry more prestige than observational research. Evidence based medicine is now supported by techniques for objectively reviewing published data (Hébert and Tugwell, 1966).

Because patients are involved and because in the past some patients have been badly treated by clinical research workers, the Declaration of Helsinki (Appendix 3) was set out in 1964, and revised in 1975 and1986, to establish the ground rules worldwide for ethical behaviour towards research on humans, whether patients or not. There is now a panoply of rules concerning research on patients and human controls, through which the individual clinical researcher will steer his way most effectively with the help of his local ethics committee, without whose approval he would be very unwise to embark on any clinical studies (Chapter 5).

Clinical trials are, of course, concerned with establishing differences between groups of patients with regard to a specified variable or variables while all other variables remain the same. This is a formidable task. There are several types: retrospective or prospective; uncontrolled or controlled. Controlled trials may be of matched pair, cross-over, group comparative or mixed design. Uncontrolled trials are those in which the potentially active treatment is not being compared with anything, and open trials are those in which both the doctors and the patients know what the treatment is. Controlled trials compare the treatment with a placebo or other treatment, and may be single blind, in which the patient does not know which treatment he is getting, or double blind, in which neither the patient nor the doctor know.

To design a clinical trial, including the randomisation procedures, has become a sophisticated exercise. Since the results will be used for various purposes, particularly for applications for the licensing of drugs for clinical use, great care has to be exercised to ensure that the design cannot be faulted.

The main application of clinical trials is in evaluating therapeutic interventions, often by drugs, by randomised controlled trials.

The number of patients needed for a trial depends on:

- the variation in the variable studied
- the minimum difference that is to be shown
- the statistical method to be used for analysis of the results
- the type I error agreed in advance
- the type II error agreed in advance

Type I error

The type I error arises when there appears to be a difference between the treatments when really there is not, i.e. the false positive. The level of the difference occurring by chance is usually set beforehand at 5% ($p<0.05$) on testing the *null* hypothesis that there is no difference.

Type II error

The type II error arises when there appears to be no difference when there really is one, i.e. the false negative. The level is set beforehand at 5% or 10% or 20% (the "power" of the trial is then said to be 95%, 90%, or 80%), by testing the *alternative* hypothesis that there is a difference.

The methods of calculating the required number of patients using preset values for type I and type II errors are set out by Lequesne and Wilhelm (1989).

Meta-analysis

Meta-analysis is the statistical synthesis of data from similar studies. This enables individually small studies to be pooled into a meaningful result. For example, of seven trials of the effects on the maturity of a baby of giving a course of corticosteroids to the mother, only two gave statistically significant results. But pooling them gave a bigger sample size and more power, and indicated that corticosteroids did reduce the risk of death from prematurity.

Meta-analysis is one of the techniques used in systematic reviews of research evidence (Chalmers and Altman, 1995). Continuously updated reviews, as exemplified by the Cochrane Database of Systematic Reviews, can shorten the time between medical research discoveries and clinical implementation of effective diagnostic or treatment strategies. Thus 33 trials of streptokinase for acute myocardial infarction produced a sample size of 36,974 patients with an impressively favourable result, although individual trials had as few as 23 patients.

The techniques of meta-analysis work best for randomised controlled trials because they are more uniform in design and have fewer biases than observational studies.The criteria for including studies must be clearly specified beforehand so that, as trials are included, there is no subsequent bias, conscious or unconscious, towards a particular outcome. By the same token all eligible studies must be included – not an easy task of identification.

Experimental research

Many clinical trials are experiments, in the sense that the researcher intervenes to alter one or more variables between the groups of patients. Experimental research is, however, usually taken to mean experiments on living organisms other than humans, or on tissues or cells which may be of human origin.

The ethics of animal experimentation are not covered by the Declaration of Helsinki. In the UK animal experimentation is regulated by the Animals (Scientific Procedures) Act 1986, which comes under the authority of the Home Office.The establishment where regulated procedures are carried out has to be covered by a Certificate of Designation. This covers any or all of scientific procedures, breeding animals and supplying animals for scientific work. The person who carries out "regulated procedures" has to obtain a Personal Licence from the Home Office and, for the programme of experimentation, has to obtain a Project Licence also. Home Office inspectors visit frequently and are very thorough. Within this framework individual laboratories have developed their own codes of practice, programmes of training and codes of ethics. Some ethics committees consider animal projects alongside human ones; others set up separate committees; while other institutions regard their procedures as adequate without an animal ethics committee. Other countries have different regulations. In the European Union, directives to cover all the member states are in preparation. In the United States, information about the laws regulating animal experimentation are on the Internet and there is a bulletin board for information about animals.

Research laboratories are also subject to stringent health and safety rules and inspection. Consent is required from the human donor of any tissues or cells, or from a competent relative if they are taken from donors who are minors or incapable of giving consent, or are dead. Thus, although the Coroner's Act 1988 empowers coroners to have parts or contents of the body submitted to examination, this can only be "with a view to ascertaining how the deceased came by his death". Thus sampling for research is not within the coroner's power to authorise. The situation varies between countries, and in the USA between those states that have a coroner system and those that have a medical examiner system. Once again the individual clinical researcher will steer his way through most effectively with the help of his local ethics committee, without whose approval he would be very unwise to embark on any experimental studies.

Experimental research has the highest prestige in the biomedical "scientific community", although the physical sciences generally have higher prestige than the biomedical sciences. Some of the most prestigious physical sciences, for example astronomy, are of course almost entirely observational! Thus the prestige game does not make a lot of sense, but may affect one's chances of funding.

Planning the work

The question to be answered or the hypothesis to be tested now has to be turned into a practical project, which will consist of making observations on some form of biomedical material, performing clinical trials on patients, or carrying out experiments on animals or *in vitro*. As Gardner and Altman (1989) have said: "There is also little help available about how to design, analyse, and write up a whole project". However, their "statistical guidelines for authors", although intended to produce statistically competent papers, can be used at the planning stage for the research as well as at the planning stage for writing the paper. Since the guidelines set out what Gardner and Altman would expect to find in the methods, it is at this stage that the research worker can start to ensure that the data referees would look for will be there.

Thus, in the methods, you must determine:

- the type of subjects, with criteria for inclusion and exclusion
- the source of the subjects and how they are selected
- the number of subjects and why that number was chosen
- the types of observation and the measurement techniques to be used

For a survey (observational study) the study design must show:

- selection of a control group, with matching procedures
- whether it is retrospective, cross-sectional, prospective
- selection of subjects and target for high participation

For a clinical trial:

- the treatment regimens, with procedures for modifying or stopping treatment
- method for allocating treatments to subjects, including randomisation
- use of blinding techniques and ensuring unbiased responses
- criteria for comparing treatments
- for crossover trials the pattern of treatments, with changeover protocol

For an experimental study:

- selection of animals or tissues
- selection of controls
- selection of experimental methods (in animals, as for clinical trials)

Statistical tests to be used:

- what comparisons are to be made

The first practical question then is: how many observations, patients or experiments will you need to get the answer? This often calls for a review of previous studies, or a new pilot study, to see what one can expect. Retrospective analysis of a group of your own cases may be useful, but analysis of published groups of cases may give a misleading impression. You want to know how big a difference in, say, blood pressure you can expect, so that you can calculate how many cases you will need for a difference of that magnitude to be statistically or clinically significant. In the case of a histological study you can calculate from the pilot study whether you will obtain a more significant result with more slides or more patients, and indeed which option entails the most work.

As all of this often calls for help from an epidemiologist and/or a statistician, this is a good time to make friends with one, because you will probably also need his help with any grant application, research and/or ethics committee approvals, and eventually with statistical analysis of the results.

Second, will the required "materials" (patients, animals, tissues) be available over your timescale?

The third step is to determine what is needed to carry out the research, in terms of manpower (can you do it all yourself?), equipment, space, technical help (clinical, laboratory, animal handling, health and safety, computer, photographic, library, clerical, etc.), time and money.

The answers to all of these questions determine whether the research should even start.

You now have the raw materials with which to draft a "plan of research" for yourself, which will be useful also for the "plan of work" section of a grant application (Chapter 5). Meanwhile, study the International Committee of Medical Journal Editors' "Uniform Requirements for Manuscripts Submitted to Biomedical Journals", and their policy statements (Appendix 1A and 1B), and the *British Medical Journal*'s additions and guidelines for referees and technical editors (Appendix 2). Science is not complete until it has been published and these sources will answer most of your technical questions on how to get into print and on to the databases!

References

Bradford Hill A, Hill ID. A short textbook of medical statistics. 12th edn. London: Hodder and Stoughton, 1991. (First published as "Principles of medical statistics". London: The Lancet, 1937).

Burkitt D. A sarcoma involving the jaws of a child. British Journal of Surgery 1958;46:218–23.

Chalmers I, Altman DG eds. Systematic reviews. London: British Medical Journal, 1995.

Cochrane AL. Effectiveness and efficiency. London: Nuffield Provincial Hospitals Trust, 1972, reprinted by the Trust with certain additions in the British Medical Journal's Memoir Club series, 1989.

Cochrane AL, Blythe M. One man's medicine. London: British Medical Journal, Memoir Club series, 1989.

Gardner MJ, Altman DG. Statistics with confidence. London: British Medical Journal, 1989.

Hawkes N. The wings of evolution. The Times Magazine, 25 May 1996; pp. 31–3.

Hébert PC, Tugwell PK. A reader's guide to the medical literature – an introduction. Postgrad Med J 1996;72:1–5.

Kerr JFR, Wyllie A, Currie AR. Apoptosis: a basic biological phenomenon with wide-ranging implication in tissue kinetics. British Journal of Cancer 1972;26:239–57.

Lequesne M, Wilhelm F. Methodology for the clinician. Basel: Eular, 1989.

Weiner J. The beak of the finch. London: Cape, 1994.

2. Literature searching and information retrieval

Gary Horrocks

Knowledge is of two kinds. We know the subject ourselves, or we know where we can find the information about it.

Dr Samuel Johnson (1709–1784)

If you are planning to fill a gap in knowledge with your research, you have to have the surrounding knowledge at your fingertips (fig. 2.1). You have to confirm that the gap has not already been filled. You have to confirm and document the knowledge that you already have. You have to make sure that there is no relevant knowledge that you do not have, and you have to have a strategy for picking up any relevant new knowledge that is published, right up to the time that you submit your own work for examination or publication.

It is inevitable that many biomedical researchers embark on their research careers from a very narrow knowledge base. Seeking out appropriate information resources to set your work in context is an important first step.

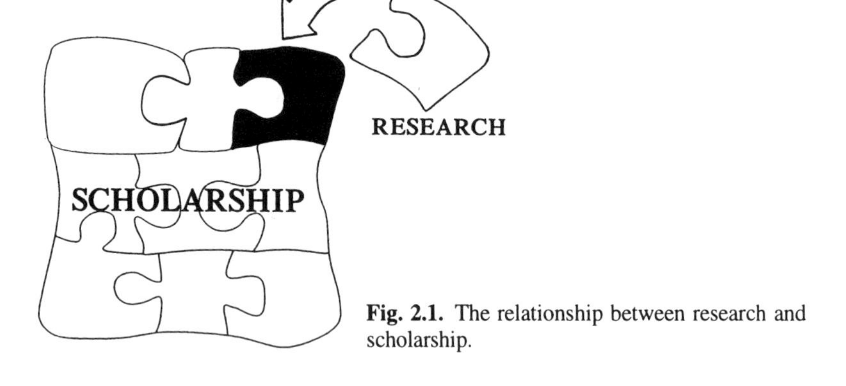

Fig. 2.1. The relationship between research and scholarship.

Understanding the tools available to help you in this task can be even more daunting.

The literature review, and the systematic process of information acquisition, organisation and exploitation, play a crucial role in the research process. They define the "state-of-the-art" knowledge in any given area, point you to related projects and highlight at an early stage whether you are duplicating projects that already exist elsewhere. A well organised literature review using all available tools at your disposal is an integral part of research methodology, and will provide the basis for your experimental work. Missing one key research paper can often devalue the impact. So where do you start?

The next three chapters highlight some of the key biomedical resources, and provide a framework on which you can build your information retrieval skills.

What you can do yourself

You need at least one starting point, a publication whose reference list leads you to other publications, often in an increasing cascade. You may already have one, from which the idea came. You rapidly build up a framework of information around your topic, one reference list or bibliography leading to another, checking each promising-looking publication in your own or the departmental or institution library for what it actually says.

Context

Biomedical research is now firmly rooted within the context of electronic networked information. The convergence of hardware, software and telecommunications technology has provided a fertile environment for the production and distribution of information. There has been a proliferation of dispersed electronic resources which has had effects not only on research workers, but also on medical practitioners and educators. Learning to access and exploit this wider range of accessible information resources will complement and add value to your research findings.

Access to on-line public access library catalogues (OPACs), CD-ROM (Compact Disc-Read Only Memory) and networked bibliographic and full text databases, alongside the spectrum of datasets and multimedia resources available on the Internet, has made research highly information intensive.

The biomedical community has to face up to the sheer bulk of these information resources. Thousands of journal articles and books in biomedicine are published every year. The National Library of Medicine in the United States currently subscribes to about 23,000 current serials. Furthermore, modes of delivering and packaging this information are forever improving and multiplying. Your library will no doubt subscribe to the New England Journal of Medicine. But is it in print form, or on microfiche or compact disc? Perhaps the full text is available to search on-line? Being aware of the potential choices awaiting you is the first step to successful retrieval.

Types of information

It is useful to distinguish between primary information, which is essentially information in its original published format, and secondary information resources, which direct you to the existing information.

Primary information

You will no doubt be familiar with most standard formal primary publications:

- books (monographs)
- journals (serials or periodicals)
- theses and dissertations
- research reports
- statistical data
- reference works (dictionaries; classification and nomenclature guides; data)
- patents

Locating a standard current textbook or monograph is a useful way to familiarise yourself with the general context and developments in your field. It provides an excellent source of references, and summarises established and accepted theories and practice. The drawback is that a book is only as current as the year it was published. It is out of date from its birth.

The academic journal is the remedy to this problem. It has been the principal vehicle of communication in science since the 1660s (Chapter 15). Current and original international research is published for the research community in many forms:

- technical reports
- critical reviews
- editorials
- letters
- randomised controlled trials

- meta-analyses of trials
- rapid (brief) communications

A review article that gives a state-of-the-art, global overview of your research activity, with an extensive bibliography, can be a launching pad to further items of potential interest. A review is both selective and critical, and will provide a useful standard of presentation for your literature review. Some journals are specifically published with review articles in mind. Look out for "Yearbooks in...", "Advances in..." and "Annual Reviews of...", for example.

Theses and dissertations are a particularly rich source of inspiration and original ideas. It is best in the first instance to look for locally published theses in your institution library. People who have gone before you may have been exploring similar themes in similar laboratory conditions. Whereas papers published in journals have gone through a process of critical assessment known as "peer review" (Chapter 15), theses are often published with a warts-and-all approach, and are more likely to be a source of more technical detail and negative results. The disadvantage of both theses and books is that they are seldom peer reviewed as thoroughly as journal articles, often to their detriment in both impact and length.

Patent information is another source of highly novel, technical and descriptive detail which may form the state of the art in any given area.

Primary information is not restricted solely to formal resources of this nature. What of informal channels of communication among colleagues? A colleague down the corridor can often be a useful source of advice and information. More important, making that first visit to your library or information service may save you hours of wasted time. Learning to ask for advice is not an admission of incompetence or failure.

Secondary information

Secondary resources (often known as indexing and abstracting publications) are signposting facilities which will direct you to original publications.

An index is an alphabetical list of, for example, authors' names or subject headings. You will be familiar with indexes at the backs of books. Journals also publish indexes and contents tables.

An abstract (Chapter 9) is a brief summary of the content of a document (a journal article, for example). It is designed to be a highly informative condensed representation of that document, reflecting the aims, methods and results of the research. It can be used to judge whether a particular study is relevant to your needs.

Indexing and abstracting publications are used extensively to locate international research as it is published in the biomedical journal literature. You can discover what is published in your area using, for example, a subject index. It will direct you to a list of references in any given area of research. A reference will usually include the minimum amount of information required to locate an item: authors, title of paper, journal title, volume, issue and page numbers. An abstracting publication will go one step further and allow you to assess the relevance of the paper before hunting it down.

Secondary resources (for example *Index Medicus*) come in multiple formats: print (sometimes referred to as "hard copy"); diskette (digitally stored on a magnetic "floppy" disk); CD-ROM (digitally stored on an optical disc like any audio CD); and on-line (digitally stored on a remote computer.) The pros and cons of some of these formats are discussed later in Chapter 3. There has been a move from standalone CD-ROM workstations to networked CD-ROM services which allow institution-wide access to these resources. Any registered staff member or student with access to the local network can usually use these resources from their desktop terminals, regardless of location. It is important that you check with your library and information service and departmental colleagues about the accessibility and cost of these crucial bibliographic services. Many services will have hybrid access to print, CD-ROM and on-line sources, so it is important that you feel comfortable with all formats. This chapter focuses on the standard electronic biomedical services on offer.

The library and information service

Make the library your first port of call. Invariably, the information services staff offer information skills training and induction courses for all incoming researchers. Get to know what your local service has to offer. How extensive is the collection in your area of interest? What training programmes are available? What abstracting and indexing services are available? Does the library have reciprocal borrowing arrangements with other libraries? The library/information officers should be viewed as friendly information resources in their own right and will be in a position to advise and demystify unfamiliar territory.

Use the on-line public access catalogues to locate primary resources. While systems will differ, they all offer standard *author*, *subject* and *title* search options to either local or dispersed library collections. There will also be browsing facilities so that you can choose, for example from alphabetical lists of subjects and journal titles. Some catalogues may be menu-driven, others may use a graphical interface such as Windows. The system used at King's College, London, offers menu-based access to the multi-disciplinary collections of the entire King's College Group of Libertas Libraries, taking in a vast multi-site collection of books and journals. The catalogues of the Institute of

Psychiatry and the United Medical and Dental Schools of Guy's and St. Thomas' are also available for searching. Further options enable the user to connect through the main menu to a number of external networked library catalogues, including the London Union List of Serials, which details the journals holdings of virtually all the colleges, schools and institutes of the University of London. Access to remote networked databases are outlined in Chapter 3.

Classification schemes exist to organise and facilitate retrieval of information in your library. The National Library of Medicine (NLM) and Library of Congress (LC) schemes are particularly popular, but your library may employ an alternative system or an in-house scheme. Familiarise yourself with those sections which may be of interest in your research area. For example, biochemistry books will be housed in the QU section of a library which uses the NLM scheme, pharmacology books under QV and immunology under QW. Don't get into the habit of browsing those areas on the shelves, as they will only lead you to what is left in the library by other borrowers at any given time, and not to what is potentially available on the catalogue.

Journals may be organised using a similar scheme, or they may be alphabetically arranged by title regardless of subject. Holdings information is usually available on-line, as most catalogues will have a periodical titles search facility. Printed lists of journals may also be on hand for reference purposes. It is important that you familiarise yourself with holdings information. Use the catalogue to check bibliographic information such as publisher and ISSN. A standard holdings statement would be something like:

Journal of Allergy and Clinical Immunology Vol. 75– 1985–[1]

This indicates that everything from Volume 75 onwards is available for consultation, but that issues prior to that will have to be located elsewhere. In most libraries, for browsing purposes, the current issue of a journal is maintained in a separate sequence from the rest.

Locating non-journal literature

There are numerous resources available to help you locate non-journal items, such as books, theses and statistics. Your best approach is to seek out advice on services in this area from your library. There are bibliographic tools, such as British Books in Print and Books in Print (US). Your information service is a gateway to a range of local, national and international library catalogues. In the United Kingdom, the British Library Network OPAC (Online Public

1 The open-ended number range is a convention used to indicate ongoing availability.

Access Catalogue) provides access for the academic and research community to a range of invaluable catalogues including the Science, Technology and Business Current Catalogue, which is a comprehensive database of international books, journals and conference proceedings from 1974. OCLC (Online Computer Library Center) in the United States offers subscription access to the WorldCat database of some 30 million bibliographic records.

In the United Kingdom the British Library is a useful source of information on theses, alongside ASLIB's "Index to Theses...". Dissertation Abstracts International is another excellent resource, particularly for North American theses.

Statistical information can be problematic to retrieve as it derives from a multiplicity of sources including official governmental bodies, international organisations like WHO and the UN, and the Internet. Your information service will advise on statistics resources and the best method of obtaining them. Often, the current journal literature will provide a useful starting point.

Accessing documents

A list of references with abstracts will be useless without the original primary source to hand. Never rely on using the abstract without referring to the original paper. It is merely an aid to relevance, and does not provide sufficient data to allow you to make a judgement on the quality, value or validity of a paper.

Use your library catalogue to locate locally held journals. Most services have extensive photocopying facilities and your department may be in a position to fund your use of these machines. It is absolutely imperative that you abide by the copyright legislation of your country with regard to acceptable limits for photocopying material from books and journals.

There will be many occasions during your research when you will require items that are not held in your home library. If yours is a multi-site institution, check out the availability of stock at other site libraries. Departments tend to maintain smaller subject-specific collections for staff, which may prove useful. In this case get to know the key players, who will advise you on the local information resources and technology available. Departmental collections often supplement the central library resources, and may contain key texts and more specialised journals.

You will generally have two options open to you: to visit an alternative local collection or to put in a request for an interlibrary loan (ILL). There may be public, hospital or special libraries with useful collections. Access to this material may be restricted, so a polite phone call asking for permission to visit may be required. If you are a member of any professional societies, use their information services. Very often they will offer a wide range of valuable alternative services. The British Medical Association's networked Medline service has some 7,000 users in the United Kingdom. If you are not a member of any such groups, find out which of them you should be joining!

If you are too busy to consult other collections, all libraries offer an ILL facility, and can obtain any items for you. Most services will charge for this facility so it is important to budget accordingly. Many countries have cooperative networks for interlending. Most types of material can be requested, from journal articles to whole journal issues, books, videos and theses. Translation facilities, often expensive and time consuming, are also available, although existing translations of a particular work may be found for you. With print items the original may be supplied to you, or a photocopy for your retention. On occasion, fiches may be supplied. Your library will be able to advise on the availability of fiche reader printers.

Increasingly libraries offer automated request input, although the procedure for manual requests is similar. You will be required to specify clearly full bibliographic information: author, title, publisher, year of publication and ISBN for books; full journal title, year, volume, issue and inclusive page numbers for journals. It is useful to emphasise your specific requirements regarding speed of delivery and format at this stage.

Your library has specialist staff trained to find the quickest and most effective route by which to obtain a requested item. Supply times will vary depending on the supplying source, the type of material requested, and its availability at the time of application. Occasionally, you may be in a queue, or have to wait for an extensive international search. Most ILL services will accept urgent requests wherein documents may be delivered by either first class mail or fax.

In the United States the National Network of Libraries of Medicine (NNLM) and the NLM provide the infrastructure for medicine, education and research, and some 4,000 hospitals, medical schools and key regional libraries cooperate in a nationwide ILL system. The NLM will only supply items as a last resort, if they are not locally available. However, its public access reading rooms provide extensive access to books, journals and audio-visual materials. The NLM has an extensive biomedical collection, with strengths in AIDS, bioethics, cancer, molecular biology, toxicology and environmental health.

The United Kingdom does not have a national library of medicine, but the British Library provides an extensive range of services. The Health Care Information Service is responsible for indexing the British journals that appear on Medline. They also provide on-line access through BLAISE-LINK to all of the NLM's databases using the Grateful Med software. An on-line search service, STM Search (Science Technology Medicine), is also available.

The Science Reference Information Service (SRIS) reading rooms in Central London have an extensive and publicly available science collection including some 25,600 current journal titles. Basic enquiry services are free, and the service can be contacted via e-mail, fax, phone or post. The Aldwych reading room houses the life sciences and biomedical collection, although the

Holborn site is useful for its collection of 36 million international patents. The British Library's Document Supply Centre (BLDSC) is a massive warehouse for international literature that serves the international research community. It supplied 4 million documents to libraries in 1994–1995, and is the world's largest source of scientific, technical and medical information.

Document delivery: electronic alternatives

Increasingly, publishers are offering electronic access to journals, either on CD-ROM or on-line. The ADONIS service, backed by a consortium of major academic publishers such as Blackwell, Elsevier and Springer, delivers the full text of nearly 700 biomedical and biotechnology journals on a weekly CD-ROM from 1991–.The full text image, including graphics, is scanned and stored on optical disc. Ovid Technologies offers a direct link from Medline to its full text collections of core biomedical journals, which include the graphics, tables and charts from some thirty journals. In the United Kingdom, the International Digital Electronic Access Library (IDEAL) from Academic Press allows licensed institutions access to the full text of almost their entire journal catalogue of over 170 journals 1995/96– over the Internet. Tables of contents and abstracts can be browsed by anyone, or the full text (including graphics) printed or downloaded using Acrobat software.

These services rely on the users' access to workstations with the appropriate software and laser printing facilities. The Internet is increasingly used as a medium for full text journal access, and your library may partake in a number of network agreements which will allow you access to such facilities. Ask advice about the availability of equipment. Your local IT department will offer advice and assistance if you run into difficulty.

[See Bibliography at the end of Chapter 4]

3. The key biomedical databases

Gary Horrocks

A database is useless unless the data can be turned into relevant information.
Madelaine Davidson (1994)

Medline

Medline, a bibliographic database from the National Library of Medicine (NLM) in the United States, is a major literature resource in the field of biomedicine. It is the electronic equivalent of *Index Medicus*, but also contains citations from the Index to Dental Literature and the International Nursing Index. It provides references to articles from over 3,700 international journals of interest to the medical profession, encompassing some seven million records from 75 countries since its inception as an electronic resource in 1966.

Like all indexing and abstracting utilities, Medline is not a "full text" database, so it does not offer whole document searching and retrieval. However, about 75% of all references are published with English abstracts, and up to a quarter of a million references to international research papers are added to Medline every year. Foreign language articles that include English abstracts printed with the text of the original article are also indicated on the database.

Although Medline's primary strength lies in clinical science, coverage is extensive, taking in chemistry and pharmacology, biological and physical sciences, psychiatry and psychology, microbiology, nutrition, health care delivery, environmental health, social science and education.

EMBASE

EMBASE is the electronic version of Elsevier's Excerpta Medica monthly series of printed indexing and abstracting publications, which started publication in 1947. EMBASE itself is not so extensive as Excerpta Medica, since electronic coverage is only available from 1974.

It is, in essence, the "European Medline", covering the international journal literature of biomedicine and pharmacology. Its strength lies in drug and chemical research, but the coverage also includes clinical and experimental medicine, biological research as related to human medicine, biotechnology (microbiology, molecular genetics, organic chemistry), health policy and management, public, occupational and environmental health, psychiatry, forensic medicine, and biomedical engineering and instrumentation.

It covers the literature from 3,500 journals from some 110 countries, and is stronger in European coverage than Medline. About 53% of the coverage is European, with a significant coverage of North American (33%) and Japanese (6%) journals. About 45% of the journals indexed for EMBASE are unique to the database, and 40% of its coverage is essentially pharmacology and drug research literature.

Like Medline, EMBASE is not a full text database, but 75% of references have English abstracts, and 75% of references are to English language papers.

Biosis Previews

Biosis Previews is a leading multidisciplinary life science database which is a useful resource for research in biology and medicine. Coverage includes immunology, pharmacology, biochemistry, microbiology, genetics, neuroscience, toxicology, biotechnology, epidemiology and public health. It covers not only basic biological research, but also field-work studies and laboratory, clinical and experimental research. It is the electronic equivalent of two printed indexing and abstracting publications:

- Biological Abstracts (1926–)
- Biological Abstracts/RRM [Reports, Reviews, Meetings] (1965–)

The online version can be accessed from 1969–, although the CD version of Biological Abstracts and Biological Abstracts/RRM is only available from 1985– and 1989– respectively.

Biosis Previews covers the literature in nearly 7,000 life science journals from over 100 countries, and processes some 540,000 items a year. Moreover, it provides an important link to publications outside the scope of journal literature:

- biology and medical books and book chapters
- proceedings of conferences, meetings and symposia (2,000 a year)
- reports

- short communications
- reviews

Proceedings are particularly useful, as research innovations tend to be aired in a conference environment a few years before they are published in the journal literature. The novelty and currency of this information will complement "formal" journal articles.

Science Citation Index Search

The Institute of Scientific Information (ISI)'s Science Citation Index (SCI) Search is the online equivalent of the printed Science Citation Index. Science Citation Index offers multidisciplinary coverage of the international journal literature, taking in about 3,300 scientific and technical titles including clinical medicine, life sciences and biology. It covers everything from physiology, molecular biology and biophysics to surgery, genetic medicine and dentistry.

ISI also publishes the popular weekly Current Contents series of literature alerting services. Its subject-based editions cover life sciences, clinical medicine, physical, chemical and earth sciences, and agriculture, biology and environmental sciences. The Life Sciences and Clinical Medicine editions are published weekly and cover about 1,350 and 990 major international journals respectively. Current Contents is a well established browsing aid wherein the research student can scan the contents pages of current science journals. There is a journal index, author index and title word index which together list journals, authors and significant words and phrases in the titles of papers. The Current Contents publications come not only in print, but on floppy disk, CD-ROM and on-line. Ask advice from your library and information service about these.

SCI Search covers about 5,200 journals in all the Current Contents fields. As biomedical research increasingly interfaces with other disciplines such as chemistry, biology and physics, this database is invaluable. It is available on-line from 1974–, and on CD-ROM (without abstracts from 1980–; with abstracts from 1991–).

ISI also produces the Index to Scientific and Technical proceedings in print, CD-ROM and on-line formats. It covers the details of papers presented at some 4,200 international conferences a year, in both books and journals.

Database essentials

There are usually two distinct approaches to information retrieval. One makes use of an artificially constructed list of subject headings used to index articles on a database. The other, known as 'free text' or 'natural language' searching, places emphasis on the terminology used by the authors to describe a subject.

Medline

The backbone of the Medline database is the MeSH (Medical Subject Headings) vocabulary, a powerful search facility which provides assistance in

Nitric Oxide
D1.625.550.5000 D1.650.550.587.600
D14.600.640

a neurotransmitter; note also specific X ref below
65
see related
Endothelium-Derived Relaxing Factor
Nitric Oxide, Endothelium-Derived see Endothelium-Derived Relaxing Factor
D24.185.188

Nitric-Oxide Synthase
D8.586.682.135.772
96; was NITRIC-OXIDE SYNTHASE (NM) 1990–95
use NITRIC-OXIDE SYNTHASE (NM) to search NITRIC-OXIDE SYNTHASE 1990–95
X NO Synthase

Nitriles
D2.626+
organic cpds having –CN radical differentiate from CYANIDES which are inorganic with –CN
68; was see under CYANIDES 1963_67
use CYANIDES to search NITRILES 1966–67
see related
Cyanides
XR Cyanides

Nitrilotracetic Acid
D2.241.81.38.600
91(73); was see under ACETIC ACIDS 1973_90

Nitrimidazine sec Nimorazole
D2.640.672.580 D3.383.374.658.580

Fig. 3.1. Part of the MeSH alphabetical list.

Term	Tree number
Amine Oxidoreductases	D8.586.682.107
Oxidoreductases, N-Demethylating	D8.586.682.107.582
Aminopyrine N-Demethylase	D8.586.682.107.582.276
Ethylmorphine-N-Demethylase	D8.586.682.107.582.400
Proline Oxidase	D8.586.682.107.640
Pyridoxaminephosphate oxidase	D8.586.682.107.680
Pyrroline Carboxylate Reductases	D8.586.682.107.695
Saccharopine Dehydrogenases	D8.586.682.107.750
Tetrahydrofolate Dehydrogenase	D8.586.682.107.825
Amino Acid Oxidoreductases	D8.586.682.135
Amine Oxidase (Copper-Containing)	D8.586.682.135.75
D-Amino-Acid Oxidase	D8.586.682.135.125
Glutamate Dehydrogenase	D8.586.682.135.398
Glutamate Dehydrogenase (NADP+)	D8.586.682.135.410
Monoamine Oxidase	D8.586.682.135.697
Benzylamine Oxidase	D8.586.682.135.697.100
Nitric-Oxide Synthase	D8.586.682.135.772
Protein-Lysine 6-Oxidase	D8.586.682.135.848

Fig. 3.2. The MeSH tree structure.

customising a search strategy by guiding you through the terminology of your field. There are two specific components to the MeSH vocabulary:

- the thesaurus, an alphabetical annotated list of subject headings (fig. 3.1)
- a hierarchically arranged "tree structure" (fig. 3.2)

The two are linked together by an alphanumeric code so that there can be easy navigation between the alphabetical and hierarchical arrangements.

In information retrieval a thesaurus works in the opposite way to a traditional Roget's vocabulary, which is essentially designed to *extend* your vocabulary by referring you to synonymous terms. A thesaurus like MeSH is often described as a "controlled" vocabulary, a list of preferred concepts designed to *contract* your choice of language and so simplify the search procedure. It is a "state-of-the-art" vocabulary which is regenerated each year to reflect current changes in the accepted terminology of the biomedical community. If a term is not on the preferred list, a "see" reference is employed to guide you to a preferred concept. A dozen or so terms from the alphabetical list are independently assigned to every article on the database, regardless of the terminology used by the authors. This eliminates the need to consider all of the various free text alternatives. It will also substantially improve the relevance of your search results because you will only be looking for items specifically assigned those MeSH headings by a specialist indexer. A free text, or 'natural language', search is potentially dangerous because it demands that you have a thorough

Table 3.1. Alphabetical list of MeSH subheadings.

Abnormalities	Legislation & Jurisprudence
Administration & Dosage	Manpower
Adverse Effects	Metabolism
Analogs & Derivatives	Methods
Agonists	Microbiology
Analysis	Mortality
Anatomy & Histology	Nursing
Antagonists & Inhibitors	Organization & Administration
Biosynthesis	Parasitology
Blood	Pathogenicity
Blood Supply	Pathology
Cerebrospinal Fluid	Pharmacokinetics
Chemical Synthesis	Pharmacology
Chemically Induced	Physiology
Chemistry	Physiopathology
Classification	Poisoning
Complications	Prevention & Control
Congenital	Psychology
Contraindications	Radiation Effects
Cytology	Radiography
Deficiency	Radionuclide Imaging
Diagnosis	Radiotherapy
Diagnostic Use	Rehabilitation
Diet Therapy	Secondary
Drug Effects	Secretion
Drug Therapy	Standards
Economics	Statistics & Numerical Data
Education	Supply & Distribution
Embryology	Surgery
Enzymology	Therapeutic Use
Epidemiology	Therapy
Ethnology	Toxicity
Etiology	Transmission
Genetics	Transplantation
Growth & Development	Trends
History	Ultrasonography
Immunology	Ultrastructure
Injuries	Urine
Innervation	Utilization
Instrumentation	Veterinary
Isolation & Purification	Virology

understanding of the terminology of your discipline. Are you aware of synonymous terms that may be referred to in the literature? Abbreviations?

Acronyms? Alternative spellings? Singular versus plural?

The tree structure will provide a contextual overview of the broader and narrower terms that exist within your chosen area. This will help you to focus the search by finding a specific aspect of a subject, or to expand the search by searching on a broader theme or on a term plus any of its more specific components, a process known as "exploding" the search.

A list of topical subheadings (qualifiers) has also been designed by the NLM to improve the specificity of a search (table 3.1). Themes range from adverse effects and diagnosis, to epidemiology and prevention and control. Any number can be linked to a MeSH heading, although the number applicable is dependent on your choice of subject heading.

Hints

- articles are always indexed under the most specific MeSH headings. If you want comprehensive retrieval of references in a given area, search on a general subject heading and use the explosion facility. For example, a search on *Tunica media* will fail to pick up information indexed using *Muscle, smooth, vascular*
- MeSH headings use North American terminology and spelling
- multi-word phrases are inverted in the alphabetical listing
- plurals are employed
- abstracts are available from 1975–

EMBASE

EMBASE offers an alternative thesaurus search facility, based closely on the MeSH headings, known as EMTREE. It consists of two volumes: an annotated alphabetical list of some 38,500 biomedical and drug related subject headings; and an indented hierarchy where the headings are organised into their subject categories. This works exactly like MeSH and is particularly useful as an aid when searching for generic drug names.

A list of EMBASE disease and drug "links" has also been constructed; these work like the Medline subheadings.

Hints

- like MeSH, articles are indexed under the most specific EMTREE headings
- EMTREE headings use North American terminology and spelling

- multi-word phrases are NOT inverted, and appear in natural word order
- singular terms are employed

Biosis Previews

Biosis Previews also uses an extensive controlled vocabulary to index records on the database. There is a "master index" of about 20,000 preferred and non-preferred terms. It includes drug affiliations, drug actions, chemical affiliations, molecular sequence data, along with notes outlining term definitions and history. Additional indexing is provided through subject or "concept" codes, taxonomic "biosystematic" codes and broad taxonomic "super taxonomic" groups.

Concept codes are codes added to a reference to represent its general theme. They are useful therefore for general retrieval of a broad subject base like medical and clinical microbiology.

Biosystematic codes are used to index organisms representing taxonomic groups above the genus level.

Super taxonomic groups are broad categories which eliminate the need to enter multiple biosystematic codes.

Medline, EMBASE and Biosis Previews all make the important distinction between major and minor concepts. Use Major MeSH and EMTREE terms and Biosis previews Concept Codes if you want to focus on the key central themes of a research document. On the other hand, minor subjects may be peripheral to the core theme of a paper, but still of importance to your research.

SCI Search

ISI's SCI Search is a natural language database that does not offer any controlled vocabulary of preferred terms to assist your retrieval, so all the pitfalls of free text searching have to be taken into account. However, it offers unique search approaches not to be found elsewhere

There is a searchable field of author-assigned keywords (arbitrarily chosen by the author on submitting the paper for publication).

The "Keywords Plus" facility lets you search on additional recurring words that appear in the titles of articles cited by the authors in their bibliographies.

The "Cited Reference Searching" facility is unique to ISI. The references cited by the authors at the end of every paper have been added to the database, and can be viewed on-line. More important, you can use them to trace the relationships between documents, and the impact a known paper has had on the biomedical community since its publication. This alternative approach can be used to augment your search strategies. If you isolate a key paper which expounds a particular theory or describes a technique in which you are interested, you can use cited reference searching to navigate forward in time from the publication of the index paper to see who else has cited it. There may be some thematic relationship between the two papers which you find of value. This is a useful tool because it enables you to search for information without being reliant on the terminological considerations of a search strategy. Moreover, citation searches may lead you to multidisciplinary papers which you would not otherwise retrieve if you stuck to the biomedical terminology. It is worth bearing in mind that citation searching will only work using the primary author of any given known paper. There is no point in trying to do a citation search using the surname of author number two in a collaborative work. Remember that citation searching is based on the concept that articles are cited because they are immediately relevant. This is not always the case. For example, as part of your literature review you may include a number of papers simply to signpost the reader to alternative areas of interest. Furthermore you may cite a paper to criticise it.

On-line versions of SCI Search offer a "Research Front" search facility. Using citation analysis the ISI identifies, and labels with a special numeric code, clusters of co-citation that occur in areas of intense research activity in a given year. Any papers that share this cluster of citations (basically all having the mini-bibliography in common) will be assigned a Research Front consisting of the numeric code and the most frequently occurring words and phrases used by the authors of the documents citing the cluster. If you find an article on the database that is of value to your research you can add depth to your search strategy by noting its Research Front number and using it to locate any other papers that share the same patterns of citation behaviour. This may retrieve useful papers that may not immediately seem relevant in terms of subject matter, but which on investigation share some fundamental point of interest, like a highly specialist laboratory technique.

CD-ROM versions offer a similar "Related Records" option, which will help you locate papers which share a high proportion of identical citations.

Which databases do you choose?

This chapter has addressed only the key biomedical databases. There are many more bibliographic services that you may have access to. Some are listed in table 3.2. It is important that you check with both your department and your library information service for access details. Each database may offer access to additional items of information, like technical reports, proceedings, books and theses, for example. If you are overwhelmed by the choice available, your

Table 3.2. Some alternative databases.

Database	Provider
AIDSLINE	NLM
BIOETHICSLINE	NLM
CAB Health – Includes communicable diseases	CAB International
CANCERLIT	NLM
CA Search – Chemical Abstracts	American Chemical Society
CINAHL – Nursing and Allied Health	CINAHL Information Systems
Cochrane Library – Reviews, synthesis and critical appraisal	BMJ Publishing
Compendex Plus (Engineering – including biotechnology)	Engineering Index, US
Current Biotechnology Abstracts	Royal Society of Chemistry, UK
Derwent Biotechnology Abstracts – including genetic engineering and cell culture	Derwent
Derwent Drug File	Derwent
INSPEC (physics, electronics and computing)	Institution of Electrical Engineers
Martindale Pharmacopoeia	The Pharmaceutical Society of Great Britain
PsychInfo – Psychological abstracts	American Psychological Association
TOXLINE	NLM

library will offer a mediated search facility, wherein relevant databases will be chosen and searched simultaneously for your convenience. Database hosts like Knight-Ridder, DIMDI and STN provide gateway access to hundreds of databases. The MEDLARS (MEDical Literature Analysis and Retrieval System) is a comprehensive system of over 40 databases offered by the NLM.

Ultimately, the key databases outlined in this chapter should be searched together, as they complement each other and will enrich the nature and scope of your literature review. While EMBASE and Medline have a clinical emphasis, the strength of Biosis Previews lies in its experimental biology and life science coverage. EMBASE's coverage is often marketed as more international in flavour than Medline, although it is deficient in areas such as nursing, dentistry, veterinary science, psychology and alternative medicine. In the field of pharmacology and drug research, however, EMBASE should always be your first port of call. The SCI Search database citation indexing offers alternative approaches to information retrieval from a vast multidisciplinary pool of science information.

A major flaw with most of these services is that they are deeply rooted in English language journal publications. Foreign language journals and non-journal materials, such as books and theses, are unfortunately not extensively covered. In this respect, Biosis Previews may be useful, as it offers the opportunity to search for books and book chapters. Advice on locating non-journal materials is given in Chapter 2 (see pp. 17–18).

Useful sources of reference to help you make informed decisions about your choice of databases are the annual lists of journals covered by each of the database producers. The NLM's List of Journals Indexed in *Index Medicus* is a useful example. It provides listings by country of publication, subject, full journal title and abbreviated journal title. Selection policy is briefly outlined. Every year new titles are added to the database after consultation with doctors, researchers, educators, editors and librarians. Others are deleted. Being aware of this element of selectivity is the crucial message at this point. Each database producer has to make informed decisions about coverage; no one database provides comprehensive retrieval of literature in a particular field. It is also useful to note that some journals are only selectively indexed if a particular paper is deemed suitable to the field of coverage. Use the journal listings as search aids to check on the coverage in your area. Is it comprehensive? Does it omit what you would consider to be essential core journals? Journal titles on Medline are input in abbreviated format, so the List of Journals is an invaluable link to full title and general bibliographic information. The List of Journals Indexed in *Index Medicus* is produced from the NLM's online journals database SERLINE, which is a publicly available Internet database of all of the serials in the NLM collection.

Your choice may also depend on the format that is available to you. The drawbacks with print publications are obvious. Manual searching is both time consuming and less reliable. Concepts have to be searched for separately, and cannot be combined by the researcher. Only major subject headings are used as access points in *Index Medicus* and the Excerpta Medical indexing and abstracting series. *Index Medicus* comes in two monthly sections, subject and author, which are voluminous and unwieldy. On the other hand, a weekly flick through a copy of the ISI's Current Contents or a check on a descriptor in the index ("three dimensional reconstruction", for example) may satisfy some people's needs for a few staple current papers!

Journal coverage may also be less for print publications. The Science Citation Index covers the literature of some 3,300 journals, but would be impossible to manage if it included the further 1,900 journals which the on-line version offers. However, print services are often more retrospective than their electronic counterparts. The Science Citation Index is available in print from

1945–, and on-line from 1974–. While *Index Medicus* itself dates from 1960–, it has predecessors dating from 1880–.

A CD-ROM database will always be less current than a regularly updated on-line service, so you may want to confer with your information service if you require highly current information. Similarly, some CD-ROM services will offer less retrospective coverage than their print counterparts.

How to approach a search

Your approach to a search will depend on your knowledge of the database and the effective use of search aids at your disposal. Let us use Medline as an example.

Define your search topic

Identify the different concepts that comprise your subject area. List them, making note of any synonyms or potentially related areas. What is your level of knowledge at this point? Are you aware of any current trends or concepts in the area you are researching? Do you already have any key papers to hand? Are you aware of key authors and organisations that might be worth noting before you begin?

```
record 1 of 1 - MEDLINE EXPRESS (R) 1/96-7/96
TI:     Adenosis (atypical adenomatous hyperplasia): hystopathology and
        relationship to carcinoma.
AU:     Epstein-JI
AD:     Johns Hopkins Medical Institutions, Department of Pathology,
        Baltimore, Maryland, USA.
SO:     Pathol-Res-Pract. 1995 Sep; 191(9): 888-98
ISSN:   0344-0338
PY:     1995
LA:     ENGLISH
CP:     GERMANY
AB:     Adenosis is a specific histologic entity which is most
        frequently seen in transurethral resection of the prostate,
        although increasingly identified on needle biopsy. The
        distinction of adenosis from low-grade adenocarcinoma is based
        on a constellation of histologic features, rather than on any
        one specific feature (i.e. nucleoli). Although mimicking low
        grade adenocarcinoma, adenosis has not been shown to be related
        to adenocarcinoma.
MESH:   *Adenocarcinoma-diagnosis; *Prostatic-Hyperplasia-diagnosis;
        *Prostatic-Neoplasma-diagnosis
TG:     Human; Male
PT:     CONSENSUS-DEVELOPMENT-CONFERENCE; JOURNAL-ARTICLE; REVIEW
AN:     96188496
UD:     9607
```

Fig. 3.3. The Medline record structure.

Define the boundaries of your search
Do you expect that a lot of information has already been published in this area? Do you want a high recall search (i.e. everything you can lay your hands on) or a precision search (a few select key papers). How far back do you want to search? Do you want to retrieve only English language papers with abstracts? Do you want review articles only? Or perhaps letters and editorials?

Search one concept at a time
This simple "building blocks" approach will enable you at a later stage to define more clearly the relationship between concepts. Use the alphabetical list of Medical Subject Headings (MeSH) to identify preferred terms and subheadings from the vocabulary constructed by the NLM to index articles on the Medline database. Use the hierarchical tree structures to place the terms within the context of their subject area. Are there broader or narrower terms that could expand or fine tune the scope of your search?

If no MeSH heading describes a theme adequately, search for words in titles and abstracts. This free text approach is particularly useful when searching for novel concepts which have yet to be adequately defined in the MeSH listing. However, it is a potential linguistic battlefield if you are not prepared for synonyms, alternative spellings, acronyms, etc. If you do not specify particular fields of information for searching, a free text search will scan the entire record and retrieve a superfluous number of out-of-context records.

Combine your search terms
Follow either the MeSH or free text approach, using the appropriate linking operator:

- use OR to link terms together representing a single concept (e.g. synonyms or related subject headings)
- use AND to combine two or more different concepts

A combination of MeSH and free text is more likely to result in comprehensive and relevant retrieval.

Refine your search
Browse through your retrieved references, and identify the articles that interest you. Familiarise yourself with the record structure (fig. 3.3). Make a note of the MeSH headings which are used to index those papers. Are there any words in the titles or abstracts that you did not use in your search strategy? Go back and add them. This interactive "harvesting" approach allows you to redefine your search boundaries continuously. It is important to be as imaginative and flexible as possible.

[See Bibliography at the end of Chapter 4]

4. The search interface and the Internet

Gary Horrocks

Having the Internet at your disposal is like having 30 million consultants on your payroll – except you don't have to pay them.

Rough Guide to the Internet & World Wide Web (1995)

The ascendancy of CD-ROM technology in the 1980s heralded the move towards intuitive, "user-friendly" software packages which provided an attractive graphic search environment and handed a degree of independence and unlimited searching to the end-user. There is a variety of commercial software packages which you may encounter during your research career. The point to be made is that they are merely a form of window dressing and that the information beneath, be it from Medline, EMBASE or SCI Search, is essentially the same. The Windows "point and click" culture has made searching the whole

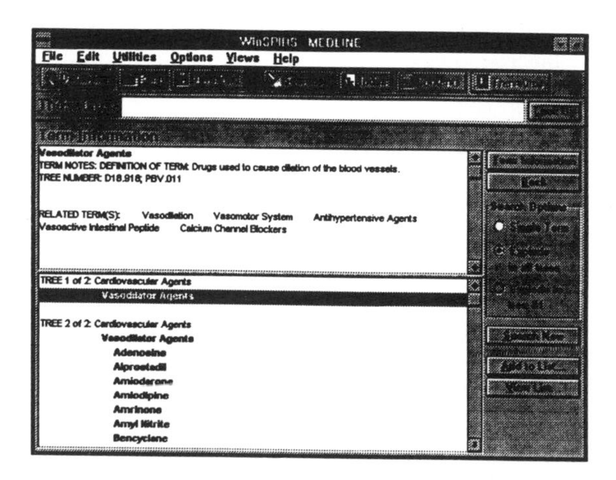

Fig. 4.1. The Windows version of SPIRS.

range of available databases deceptively easy. However, understanding the construction of and indexing and selection procedure for any given database is far more important than grasping the mechanics of each information retrieval software package you encounter. Invariably, local documentation, help sheets and on-line help screens will be provided. Once you have mastered the theory and technique of search strategy formulation, an unfamiliar screen and different keyboard conventions will be easily overcome.

SilverPlatter's Information Retrieval Software (SPIRS) comes in three configurations, DOS (PCSPIRS), Windows (WinSPIRS, fig. 4.1) and Macintosh (MacSPIRS). Ovid Technology's OVID interface is also a market leader. Most packages offer common search facilities with which you should familiarise yourself. These include:

- truncation and wildcard searching, e.g. counsel*, p?ediatric
- use of the logical operators AND and OR
- field-specific searching, e.g. words such as "nitric oxide" in titles or abstracts
- customised record display formats
- marking, downloading and printing of results
- saving search strategy for reloading
- online thesauri
- explosion searching
- natural language to thesaurus mapping
- index browsing facility
- limit options, e.g. by publication year, language, publication type

The National Library of Medicine's Grateful Med software is a popular search interface for the NLM range of databases and has sold about 60,000 copies to date. It enables the user to browse the MeSH headings and formulate a search strategy off-line. It then automatically runs it and downloads the retrieved records. Another popular interface to the Medline database in the US is the PaperChase service.

Early 1996 saw the fifth anniversary of the UK BIDS initiative. The BIDS (Bath Information and Data Services) Gateway from the University of Bath provides network access across the UK research and academic network (JANET – Joint Academic NETwork) to key biomedical, life sciences and drug literature, including EMBASE (1980–) and SCI Search (1981–). Subscribing institutions in the academic and research sector are able to offer free access to registered staff and students. Peaks of up to 9,000 users a day have been recorded for the ISI service.

Weekly database updates and the relatively user-friendly menu-driven interface (with a parallel graphic Internet interface expected towards the end of 1996) has made BIDS an important information retrieval service in the academic sector. Moreover, the service is available 24 hours a day with only the occasional downtime for maintaining and updating the service.

As part of the general tendency towards consolidating the academic and research communities, the Edinburgh Data and Information Access (EDINA) gateway was launched in the UK at the end of 1995. Subscribing institutions in the UK academic and research community offer free access to Biosis Previews from 1985– to registered students and staff. Ask for advice about the availability of these important services. If you are eligible for access you will be asked to fill in a copyright declaration form, and will be issued with a user name and password.

The Internet

No one can have escaped the onslaught of media attention surrounding the Internet. It has been labelled, among other things, the "network of networks", the foundation of the "Information Superhighway", "Cyberspace" or simply "the Net". It is, in essence, a set of international, interconnected computer sites, from the lowliest PC to the largest mainframe computer; a conglomeration of networks – research, governmental, commercial and educational. In this respect it comprises a huge global information resource. In the UK, for example, JANET, the Joint Academic Network, has played a substantial role in contributing to this global resource and has traditionally offered a number of valuable information services to the academic and research communities. This chapter can only address some broad themes which may be of value to the research student. There is a range of publications available for you to consult which address in detail established network applications. Your information service will be able to offer training, advice and assistance to enable you to exploit the Internet as an information tool.

What does the Internet have to offer?

The Internet provides continually improving access to a massive and expanding pool of information including international biomedical and clinical resources. Not only does it link you to "traditional" bibliographic information services (such as library catalogues; on-line databases such as Medline), but it offers through the World Wide Web (WWW) a varied medium for the retrieval of audio-visual media and textual and graphical documentation from a range of organisations including universities, research establishments and hospitals. These resources are often more up to date than their printed counterparts (if any exist), and can be accessed at the touch of a button or the click of a mouse.

Moreover, many such services are offered free of charge, and the information can be downloaded and used locally.

Some examples:

- electronic mail: communication and research collaboration
- document delivery: electronically accessible newsletters and journals
- software delivery: public domain software packages, from biomedical computer assisted learning packages (CAL) to whole downloadable genome databases
- access to bibliographic and full-text databases
- access to image databases
- access to current awareness and literature alerting services
- access to worldwide library catalogues and Campus Wide Information Services (CWIS)
- electronic discussion lists
- resource guides and directories

Electronic mail (e-mail)

This enables the worldwide exchange of messages from one computer to another without the confines and delays of traditional mail. It has fostered a rapid and simple means of communication and the exchange of ideas. The informal dissemination of prepublished information has made it a crucial medium in the research process. Job notices and research vacancies can be posted to a wide audience. Immediate responses can be made to published articles with a quick comment to the author or editor. Questions can be answered and laboratory techniques compared. You can use your e-mail utility to subscribe to a whole range of discussion lists in your subject area and participate in the exchange of ideas with a like-minded group. Electronic mail has further blurred the distinction between formal and informal communication in science.

Electronic journals and document delivery

The Internet has offered up a whole new educational space for the delivery and dissemination of research information. The "electronic journal", and the dynamic potential for interaction and communication that it has made possible, have stimulated a major area of network experimentation and activity. The electronic journal as such may range from the parallel digital version of a print copy, to a purely electronic presence. It may be a newsletter, a magazine or a cumulation of discussion list activity. Core biomedical journals offer a network presence which may consist of table of contents listings, abstracts, job advertisements, general news and instructions to authors. Information will often be available to you which has not yet been added to the key biomedical databases. While some may offer full text access, please note that most of these titles are commercial concerns, and will ultimately require a subscription payment to access the full service. Check with your information service for accessibility details. Your institution may have signed a license agreement with the publisher which allows full text access from your desk top. Most database producers like Elsevier and the ISI offer commercial services allowing you to place direct orders with them for the full text delivery of items. It is best that you discuss document delivery options with your information service.

Access to bibliographic services

The WWW has revolutionised access to all the services so far outlined in this chapter. Web forms-based interfaces are now available from both SilverPlatter (WebSPIRS) and Ovid's WWW gateway. The Academic Reference Centre (ARC) service in the UK offers Internet access to multiple academic databases using the SPIRS software. In April 1996 the NLM launched its Internet Grateful Med (IGM) service. It offers assisted searching of databases like Medline over the WWW. The Knight-Ridder company has also launched KR ScienceBase which offers seamless access over the WWW to databases like MEDLINE, EMBASE and SCI Search. The point to be made is that, as these services proliferate, it will become increasingly difficult to make informed decisions about availability and access. Ask for advice.

Current awareness services

A whole range of literature alerting services is available on the Internet. Publishers' home pages offer access to their journals' contents pages. Elsevier Science's TOC (tables of contents) service is one example. The Uncover service, available to search on the WWW, is a free database of some 17,000 multidisciplinary journals, about half of which are in science and technology. However, to set up a customised alerting service where the tables of contents of the key fifty journals in your field can be e-mailed to you on publication will currently cost you $20. There is also a charge for fax delivery of the full text of any items you require.

Library catalogues

The Internet enables access to hundreds of library catalogues and public domain databases. Examples include: the NLM's Locator database of book, journal and audio-visual holdings; those of the British Medical Association; the Library of Congress catalogue. The UK Wellcome Centre for Medical Science offers the WISDOM service, which includes a database of biomedical research assistant vacancies, alongside another database specifically addressing sources of biomedical research funding.

Some considerations before exploring the Internet

Anarchy

In *The Whole Internet*, Ed Kroll (1994) likened network exploration to handling jelly. "[T]he more firm you think your grasp is, the more [it] oozes down your arm". The problem that confronts the bewildered user lies in the very nature of information on the Internet. Like a library without a catalogue it is anarchic, and there is no formal organisation or quality control. Any one resource may be accessible in a variety of formats, and there are multiple search tools and routes to access this information. You are able to find books and journals in the library because they have been meticulously catalogued and assigned appropriate subject headings. No such organisation exists on the Internet, although there are a number of initiatives under way to resolve this situation.

Quality

There is a lack of regulation and peer-review of resources. This brings into question the accuracy and comprehensiveness of information. A lot of trash is available for consumption, which will necessitate a greater degree of selectivity and evaluation. In turn, quality services are heavily used and often difficult to access.

Currency

The Internet is in a constant state of flux. Site addresses continually change, and some sites are neglected and cease to be updated to the point where they are rendered useless. Look out for dates and update information.

Is it all free?

While there is a significant body of high-quality public domain information available, value-added services like the Medline and EMBASE databases will require some form of user ID and payment. Occasionally you will come across "free" versions, but they invariably offer limited coverage and rudimentary interfaces which severely restrict your search options.

Subject directories and search engines

There are numerous projects and initiatives under way to catalogue and index Internet resources, such as the WWW Virtual Library, Yahoo and Health Web. The US Medical Matrix is a hypertext listing of sites and services which categorises information by disease, specialities, interest areas, quick reference guides, image databases, learning modules and interactive forums. OMNI (Organising Medical Networked Information) is a UK gateway to quality filtered resources in biomedicine. The majority of these services are, in essence, browsing facilities. Some also provide keyword search options.

Search engines are software initiatives designed to search databases at WWW sites. Key services are Lycos, Alta Vista, WebCrawler and World-Wide Web Worm. At regular intervals the Internet is dredged for new sites, which are then added to the service's catalogue. The scenario is deceptively simple: all you have to do is input the appropriate keywords, and the system will go off and locate any matching sites. However, all search engines should be handled with care. There is no one ultimate search tool, and no one comprehensive database of WWW sites. Each service offers different search options. Look out for on-line help. How are search queries to be constructed? Does the service offer the opportunity to limit the number of items received, so preventing an inundation of retrieved items?

Search hints: a general approach to information retrieval on the Internet

- Ask advice. Your library may be able to advise on a quicker, more efficient method of locating information. There is no point losing yourself on the network when a reference book is to hand locally. Similarly, a colleague in your department could have information of value.
- Isolate the appropriate discussion lists of interest – they are invaluable sources of information. Use the resource browsing facilities and search engines; often one reliable site in your research area will be all you need to

link you to other related resources.

- Approach your network exploration with the same care you would take over a Medline search. Take note of synonymous terms, acronyms, plurals and alternative spellings. If in doubt, use logical operators. Some systems assume an AND operator, others assume an OR.
- WWW browsing software packages like Netscape have "hotlinks" or bookmark facilities so that you can build your own library of locations of value. Use it.
- Try to search local sites if there are multiple access points available.
- Try to search at times when you think the rest of the world will be in bed, otherwise excess traffic will considerably slow things down.
- If a service is temporarily unavailable, try again later.
- Know when to stop.
- Last, but not least, the Internet is a complement to other resources, not a replacement.

Molecular biology on the Internet: a quick case study

Research in the area of molecular biology and genetic mapping is fundamental to the biomedical community. International researchers have used e-mail, bibliographic databases and genome and sequencing datasets to gain access to crucial current information which will define the direction of research for years to come. E-mail is used extensively for the retrieval and analysis of genetic data. There are programs for sequence similarity searching (e.g. BLAST). The GENBANK database of genetic sequences is publicly available for searching. Online Mendelian Inheritance in Man (OMIM), the catalogue of human genes and genetic disorders, has an Internet presence. Discussion lists, newsgroups and resource lists for molecular biologists have stimulated the dissemination and sharing of resources on a large scale.

Managing the information

It is important to maintain accurate and complete information about all of the items that you have consulted. One still comes across many references by other than electronic routes and it is imperative to capture the full reference on first acquaintance, to avoid finding oneself fruitlessly trying to quote references known only as, for example, *Smith? 1977? American Journal of Pathology?* Recognising the problem early on, and tackling it with self-discipline, are the only answers to this irritating phenomenon. The manual approach must thus be combined with the electronic approach. This is gradually becoming true worldwide; even those researchers working in low-tech circumstances are progressively being included in the information revolution, and should take serious note of the tools outlined in this chapter.

The most important rule is to be consistent when recording information. If you have come across a reference by manual means, ask yourself some questions. Have you noted down all of the authors with their correct initials? Have you made a note of the title of the paper, year of publication, volume, issue and inclusive page numbers? Learn to annotate your records with useful hints and reminders of the value of the paper. Note the location of the full text so that you can easily consult it at a later stage.

The proliferation of electronic information resources has made it increasingly difficult to manage and control citations. Fortunately, the age of record cards and hastily scribbled references is declining, and there are software packages available specifically designed to manage your references. Personal bibliographic packages like Reference Manager, Papyrus, Pro-Cite and Endnote are becoming essential research tools. They enable you to store, retrieve and edit references, prepare bibliographies and effortlessly incorporate them into the word-processed body of your papers. Citations downloaded from electronic databases like Medline and EMBASE can be imported into your own personal database and then manipulated accordingly. Multiple styles of bibliography can be generated to suit your needs, and all kinds of information types, from journal articles to book chapters, can be included. Your information service will be able to advise on the availability and use of such packages.

Keeping up to date

It is important that, once you have mastered search methodology, you continue to keep abreast of current information and developments. Most systems enable you to save and re-run your search history. Get to know how to use these facilities, and incorporate them into your search procedure. This will avoid any unnecessary duplication of effort and allow you to set up effectively your own current literature alerting service by regularly running your search against the latest month's or week's data. Get to know what alerting services are available on the Internet.

Your library and information unit may offer current awareness bulletins in a variety of formats, including (usually for a fee) both print and online table of contents services. They will inform you at regular intervals of the latest articles in your research area. Keep a look out for library information sheets and newsletters which may alert you to a new information service.

Your department may have a journal circulation procedure. Don't rely too heavily on this, as issues tend to lie forgotten in people's offices or laboratories. If you come across a useful paper, bring it to the attention of your colleagues, who may also benefit from reading it. Complementing your formal information retrieval skills with informal communication and cooperation among colleagues is ultimately the most effective approach to your research.

Conclusions

Biomedical information in all its guises, from print to CD-ROM and the Internet, is complex and diverse. The onus will always be on you to remain aware of the vast pool of potentially useful resources at your fingertips. Your research will be flawed and inadequate without a comprehensive approach to information retrieval. It will provide not only the basis of your literature review, but will inspire and inform you throughout your career.

As Jenkins (1985) concluded: "Remember that in this final step, that of publishing, you will not only be communicating the results of your work to the scientific world but also adding to the ever-increasing scientific literature which will have to be searched by those who come after you."

Bibliography

Davidson M. Computing and information Technology. London: Straightforward Publishing Ltd., 1994.

Glowniak JV, Bushway MK. Computer Networks as a Medical Resource: Accessing and Using the Internet. Journal of the American Medical Association 1994; 271:1934–9.

Jenkins SR. Searching the literature. In: Hawkins and Sorgi (eds). Research: how to plan speak and write about it. London: Springer-Verlag, 1985.

Kennedy AJ. The Internet & World Wide Web. London: Rough Guides Ltd., 1995.

Kroll E. The Whole Internet: User's Guide and Catalog. Sebastopol, O'Reilly and Associates, 1994.

Lee R. How to Find Information – Life Sciences: a Guide to Searching in Published Sources. London: British Library, 1992.

Lowe HJ, Barnett GO. Understanding and Using the Medical Subject Headings (MeSH) Vocabulary to Perform Literature Searches. Journal of the American Medical Association 1994;271:1103–8.

McKenzie B. Medicine and the Internet: Introducing Online Resources and Terminology. Oxford: Oxford University Press, 1996.

Morton LT, Godbolt S. Information Sources in the Medical Sciences (4th edn). London: Bowker Saur, 1992.

Morton LT, Wright DJ. How to Use a Medical Library (7th edn). London: Library Association, 1992.

Woodsmall RM, Benson DA. Information Resources at the National Center for Biotechnology Information. Bulletin of the Medical Library Association 1993;81:282–4.

5. Applying for ethics committee approval and research grants

Ethics and science need to shake hands.
Richard C Cabot (1868–1939)

Remember the end never really justifies the meanness.
Anonymous American

If the plans for your research include applying for funds, the application form will ask if you have ethical approval for it. To get ethical approval you will have to apply to the Ethics Committee of the institution in which you plan to do the research. Consideration of ethical approval therefore comes before the grant application.

The formats of the two application forms differ somewhat from institution to institution and from country to country, but the information the bodies require has become fairly uniform. The current forms at King's College Hospital and Medical School, shown in Appendix 4, are a fair sample of such forms, the grant application form being based on that issued by the Medical Research Council (MRC). Completing these application forms is hard work because you have to think of clear and convincing answers to what you think might happen. But, as with the planning, this will greatly improve the work you do and the clarity with which you write it up. At least in the case of the latter you know what *has* happened.

Ethics committee approval

Research is bound by ethical considerations in relation to the way patients or people are treated. This is a complicated subject that has and continues to evolve. Suffice it to say that it is not now acceptable for anyone to carry out biomedical research on their own initiative without scrutiny and approval having been given by the local ethics committee, which consists of a wide range of members including lay members. The sanction for failure to observe the ethical requirements is that the journals will not publish the results without confirmation that the project had been approved, and even then they may judge that the work was unethical.

The form (Appendix 4) will cover the selection of subjects, the information to be given to them and their general practitioners, their informed consent, and what will be administered to them or done to them, as well as the objective, design and scientific background of the research, including what has previously been done on animals or humans and why more should be justified.

Sources of funds

There are many more sources of funding than the researcher will expect. They range from international bodies such as the European Union (EU) in Brussels, multinational pharmaceutical and other industrial companies, overseas foundations and government bodies such as the National Institutes of Health (NIH) in the USA, Max Planck Institutes in Germany, Ludwig Institutes worldwide; and from the national research councils, such as the Medical Research Council (MRC), and charities, such as the Wellcome Foundation, through a multitude of smaller research and charitable organisations, often set up to tackle research into specific diseases or biomedical problems. Many hospitals have their own charitable research trusts and many individuals have set up small charities, sometimes centuries ago, to which researchers can apply for funding. Charitable status is important in the UK because it exempts the organisation from taxation on its income, and purchases made with its funds are exempt from Value Added Tax (VAT). It is quite easy and cheap to set up a charity oneself and to register it with the Charity Commission[1] for charitable status, although this has been tightened up in recent years. Other countries have similar arrangements to explore. Advice about funding bodies is usually available within the researcher's organisation, and from the Charities Aid Foundation[2] which produces a useful book of charities. In addition the National Health Service has a research and development division, with national and regional directors, to fund research and development of interest to the service. Funding bodies, including societies and associations such as the British Medical Association (BMA), frequently advertise for applications for research grants in the biomedical and scientific journals, often for those small sums for equipment or travel that are otherwise difficult to fund.

1 Charity Commission, 57 Haymarket, London SW1Y 4QX

2 Charities Aid Foundation, 48 Pembury Road, Tonbridge, Kent TN9 2JD

Constructing the application

The forms that the funding bodies supply often follow the MRC format, as shown in Appendix 4. There are usually at least ten components, including the names of the lead researcher and co-workers and/or supervisor. One has to give a *title*, sometimes not to exceed a specified number of characters (including spaces), an *abstract* of specified length, the *purpose* of the proposed investigation, details of the *funding* requested, in the categories of staff (if necessary including a salary for yourself), equipment and consumables. Then you have to write a *background* to the project, with a review of the relevant literature, your *plan* of research (with references to your previous work, if appropriate), and your *reasons* for seeking funding. The final items include the curriculum vitae of each applicant, the approval of the director of the service in which the work is to be carried out, of the accountant to confirm that the financial aspects have been correctly calculated, and of the ethics committee if appropriate. There may be other criteria to conform to; for example, the EU applications usually require collaborators in one or more of the other member states, and an indication as to the potential industrial or commercial value of the work.

Obviously this involves a number of man-hours of work to collect, write, present and obtain the approval of your co-workers; make sure you don't get in a muddle with different drafts of the submission. Do not, however, forget that it also means man-hours of work for the assessors, referees and committees who decide which applications to fund. Anything which makes it easier for them is a point in your favour. In particular, do what the form asks. Keep within the space provided. Make the presentation visually pleasing and easy to read and assimilate, at least by having it typed, or printed from your word-processor. If the latter, use the typeface and size specified, though varying the size of type, if it is not specified, may help to keep within the space.

Look at the application from the committee members' point of view rather than your own. They may have many applications to read, so keep yours simple. They have written such applications themselves, so they will not be blinded by science, and in any case they may well have the science judged by referees. They are also interested in your track record. Have you had grants and carried research through to completion before? In this field? Was it published in prestigious journals (i.e. have other judges passed your work?) Have you a team, or are you a one-man-band? Is the topic exciting, important, fashionable? Is it in the field of the grant-awarding body?

You can be imaginative, for example by including an illustration or a figure to make the method easier to understand. Send the number of copies asked for, and construct a careful covering letter, explaining why this particular body is the one to fund your project to the credit of itself as well as you. You may even suggest who might be appropriate to referee it or at least point out that it is in conflict with the ideas of someone they might think of sending it to. Get it in by the closing date. Having done all this work will help you tremendously when you come to write up the research for presentation and/or publication.

As always in the biomedical field, prestige is important, however unscientific it may be in its assessment and application. So many researchers become depressed when applications to the Wellcome or the MRC or the EU are unsuccessful after months of hopeful waiting and worrying about whether you will be able to keep your team together. Nevertheless, with a well constructed application, the researcher can usually get funding from somewhere, eventually.

B
RESULTS

6. Data handling and interpretation

"Science is Measurement"
by Henry Stacy Marks

The statistician in medicine should be regarded as an obstetrician rather than as a morbid anatomist.
Dr FHK Green, quoted by
Sir Austin Bradford Hill (1946)

No human investigation can be called true science without passing through mathematical tests.
Leonardo da Vinci (1452–1519)

Biomedical data tend to be very untidy, a mixture of textual information about patients, animals or specimens (let us call them "cases" for simplicity) and numerical results. "Data" (Latin) itself is an awkward word; as the plural (of "datum") it should take plural forms of verbs, and it does so in this book, although this sometimes appears clumsy. Since many people use "data" as a single noun, and this is quoted in the Shorter Oxford Dictionary, perhaps the solution would be to judge your readers' preference. Whatever you decide, be consistent.

Primary data come in three forms. If the data are in mutually exclusive categories, for example male/female, positive/negative, present/absent or American/European/Asian/African/Australasian, with no particular ranking, these are *nominal* data. If the categories are mutually exclusive but have some form of order or ranking, such as social class (I to V), or grade of malignancy (+, ++, +++), they are *ordinal*.

Numerical data, such as body weight or age, are *quantitative* data, which can be discrete or continuous although the boundary between them is fuzzy. Discrete data are limited by their nature in that, for example, the number of patients must be a whole number. Continuous data usually result from measuring something, to however many decimal points the researcher may require, for example body weight.

When examining the primary data, those for each patient may be examined singly: for example, the distribution of body weight. This is *univariate* analysis, and may be set out for examination as a linear "dotplot", bar chart, pie chart, or frequency distribution (Chapter 7). If two variables are examined for each patient, for example body weight and blood pressure, this is *bivariate* analysis. The variables may, of course, be nominal, ordinal or quantitative (discrete or continuous), or combinations of these, of which there are ten (Coggon, 1995). There are many ways of displaying these combinations, most commonly as graphs with one variable on the x axis and one on the y axis. It is more difficult to set out multiple variables and examine them by *multivariate* analysis.

Frequency distributions are commonly used to summarise quantitative data and in univariate analysis of quantitative data. The summary concentrates on measures of the middle of the data, such as the mean, median and mode; and on the spread, range or dispersion of the data, with, for statistical testing, the wild or outlying data points.

The frequency, for example the number of patients in each of a number of categories of blood pressure reading, may be plotted as a histogram (Chapter 7) and a curve drawn to link the categories. The purists do not like such curves because the categories are mutually exclusive and not linked, but as long as the reader bears this in mind they are a useful summarising device.

If the curve turns out to be a symmetrical, bell shaped curve, it is described as a "normal" curve. The central summary measures, mean, median and mode, all coincide (fig 6.1). The arithmetic mean or average is the sum of the readings divided by the number of readings, a minor manipulation of the data. The median is the reading in the middle of the whole range of readings. The mode

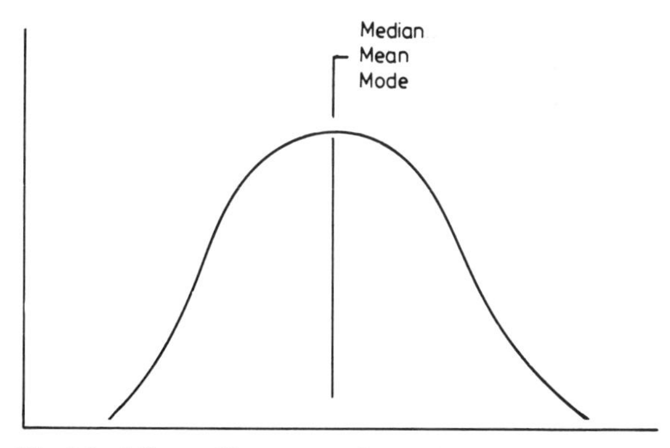

Fig. 6.1. A "normal" curve; one in which the mean, median and mode all coincide.

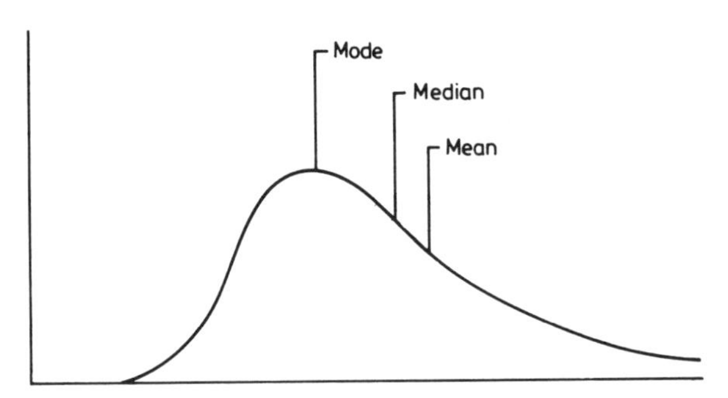

Fig. 6.2. An asymmetric, or "skewed" curve, in which the summary measures do not coincide.

is the reading with the greatest frequency, i.e. the most popular. On the other hand, if the curve is asymmetric or "skewed", the mean, median and mode usually do not coincide (fig. 6.2). The observer may easily explain such asymmetry, but the importance of knowing that it exists within the data is that it affects the statistical tests that may be applied to show probabilities, hypothesis testing or confidence intervals (see pp. 59, 62–3).

A summary of the dispersion is shown by the range, which is sensitive to any outlying readings. The data can be manipulated to give the variance and the standard deviation (SD) of the data set (see p. 56). It is a useful statistical fact that 68% of the cases will fall within the range of the mean plus or minus SD, while 95% of the cases will fall within the range of the mean plus or minus SD x 1.96 (or x 2 for practical purposes) (fig. 6.3a and b).

One hopes that, at the planning stage (Chapter 1), very careful thought went, first, into deciding exactly what primary data to collect into each "casebook", and into the forms – usually on paper but increasingly electronic – on to which the items of information were to be collected. Second, the computerised layout of each casebook should have been carefully determined, not only for ease of keying-in the data from the forms (if they were not primarily electronic), but also for sorting and manipulating items from the whole set of casebooks with minimal additional keying. This may need advice from a computer-experienced colleague.

Care must be taken to retain dated original forms or electronic entries (with dates) indefinitely, so that they can be produced to establish the authenticity of the data if there is ever any question about plagiarism or forgery, or even if there is a question of precedence. The same applies to laboratory notebooks containing ideas, methods tried, modifications to equipment; and to day books containing specimen accession and tests done, even in routine laboratories and operating theatres; and also to case notes. Photocopies of those not releasable to you may have to suffice. Biomedical research workers are notoriously care-

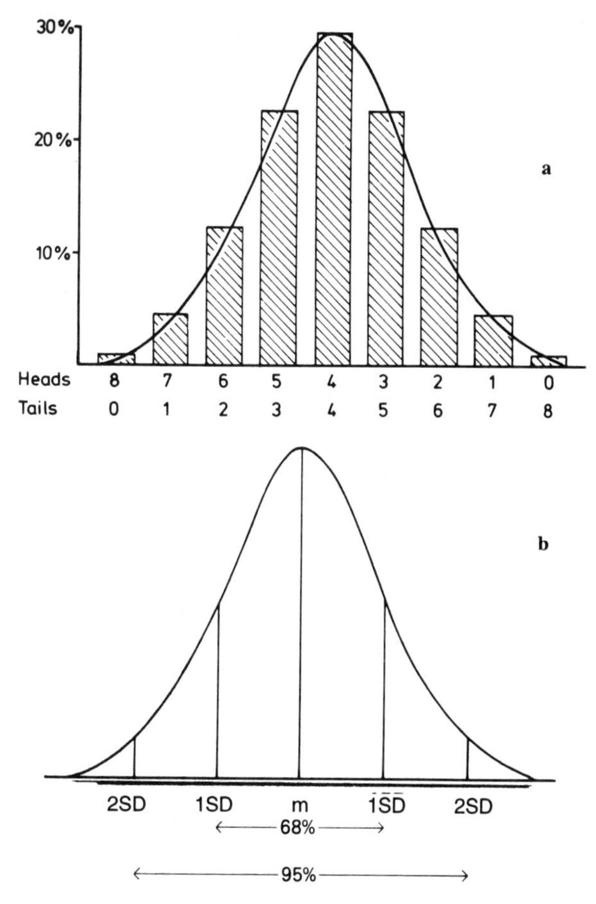

Fig. 6.3. Curve **a** shows the frequency distribution; curve **b** shows the standard deviation (SD).

less about recording and keeping their primary data after they have used it; it is no longer safe to be careless (Lock and Wells, 1993).

Raw data

Except sometimes in case reports, with one casebook, the primary or raw data do not usually stand on their own. In most instances data from one series of

Table 6.1. Raw data, listed for both antibiotic and placebo groups.

Boil Trial
Recovery time: hours

Case number	Antibiotic	No treatment	
1	40	54	
2	43	52	
3	52	51	
4	49	52	
5	41	49	
6	58	51	
7	55	56	
8	41	60	
9	43	58	
10	44	53	
11	48	49	
12	54	50	
13	55	50	
14	51	60	
15	48	60	
16	47	48	
17	49	53	
18	50	56	
19	48	47	
20	48	44	
mean	48.2	52.6	(diff = 4.4)
SD	5.09	4.53	
SE	1.1	1.0	(sum = 1.15)
Range	40–58	44–60	

SE of difference 1.52 **t = 2.89**
degrees of freedom 38 **P<0.01**

90% confidence interval **–6.97 to –1.83**
95% confidence interval **–7.48 to –1.32**
99% confidence interval **–8.53 to –0.27**

If sample size were 100 in each group
95% confidence interval –5.74 to –3.06
If sample size were 1000 in each group
95% confidence interval –4.82 to –3.98

casebooks have to be compared with data from another series. As we shall see, there are no simple examples, but perhaps a fairly simple example would be the clinical trial of an antibiotic which is given to one group of patients, each suffering from a boil (caused by *Staphylococcus aureus*), while a placebo is given to another group of patients. We want the only variable to be the administration of the antibiotic, so in the planning stage thought had to be given to patient or microbiological variables that might affect the natural history of the boil, such as concomitant illness, congenital susceptibility to bacterial infection, or differing subtypes of *S. aureus*, and which might differ between the groups (confounding factors). We believe, therefore, that our exclusion criteria were comprehensive and that our inclusion procedure (randomisation) gave each patient an equal chance of being given the antibiotic or the placebo. In this experiment the question is one of time: does the boil get better more quickly in the antibiotic group or in the placebo group?

We first extract the time from antibiotic administration to healing of the boil from each casebook. Time is a continuous quantitative form of data, which we decide to handle as discrete hours, taking the time to recovery to the nearest hour. We then list these raw data separately for the antibiotic and placebo groups (table 6.1). This gives the two ranges of times, which may or may not overlap. Even if they do not overlap, the distributions are not easy to comprehend from the lists, so we need to look at them graphically, in a univariate way (one variable per patient). Thus fig. 6.4 shows that the results are nicely grouped with no outliers to skew the distributions. Outliers do not affect the median, but they may make the mean unrepresentative. The normality or

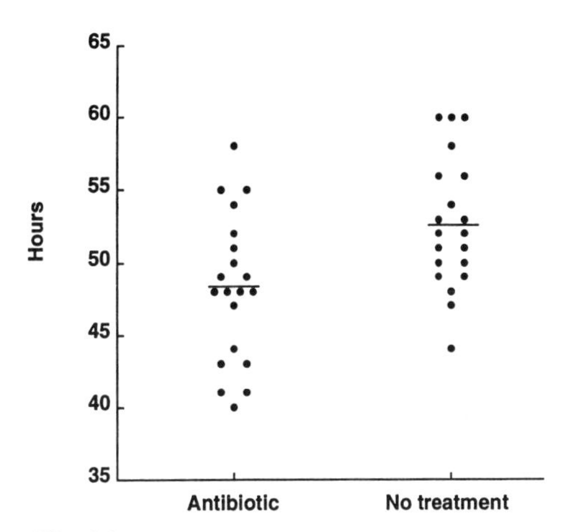

Fig. 6.4. A graphic representation of the distributions shown in Table 6.1.

skewness of the distributions can be formally tested. If they are normally distributed we will use parametric statistical tests; if they are skewed we will use non-parametric tests.

The raw data are important because they give the research workers, their inevitably less intimately involved supervisors and their uninvolved readers an overall impression or *feel* for the value of the results of the groups being compared, in terms of the size and uniformity of the groups, the homogeneity and spread of the results and the size and credibility of the differences between the groups. Naturally, all these aspects can be examined by applying statistical tests, but a common-sense comprehension of the raw data is a good guide to the initial interpretation of the biomedical, as opposed to the statistical, significance of the outcome of the experiment. Unfortunately, many experiments produce too much raw data to publish except in theses, so you must be selective, not to show merely the results that favour your hypothesis, but to give the results that allow the reader to judge the basic features listed above. Some editorial offices will allow raw data to be filed with them so that readers can be sent them on request. It is also possible to make raw data available to readers electronically, even through the Internet.

Manipulated data or data description

"How can we summarise small amounts of data?" asks Hutchinson (1995). Data may be made manageable by manipulating them, often by summarising the results from groups of cases into representative or summary statistics, such as the *mean* and the *median*, and an indication of spread or variation, such as the *variance* (the average of the squares of the differences of each case from the mean), the *standard deviation* (SD: the square root of the variance), or the *standard error* of the mean (SE: the SD divided by the square root of the number of cases). The SE is an estimate of the spread of means that would be obtained if many groups of cases were given the antibiotic in the above example. Furthermore, 95% of means from many groups would fall within the range of the mean plus or minus SE x 1.96. As a rule of thumb, if the means of the antibiotic group and the placebo group are separated by more than the SE of the first group plus the SE of the second group, the difference between the means is likely to be statistically significant at the 5% level.

However, summary statistics alone are not enough for your supervisor, editor, referee or journal reader to assess the validity of your data. Readers also have to know the number of cases, the range of the results, together with outliers and normal or skewed distribution; i.e. they need some raw data too. That this is often lacking in published papers is a comment on the journal. What is worse is a mean given as follows: 26mg ± 2.5. The ± may come from the

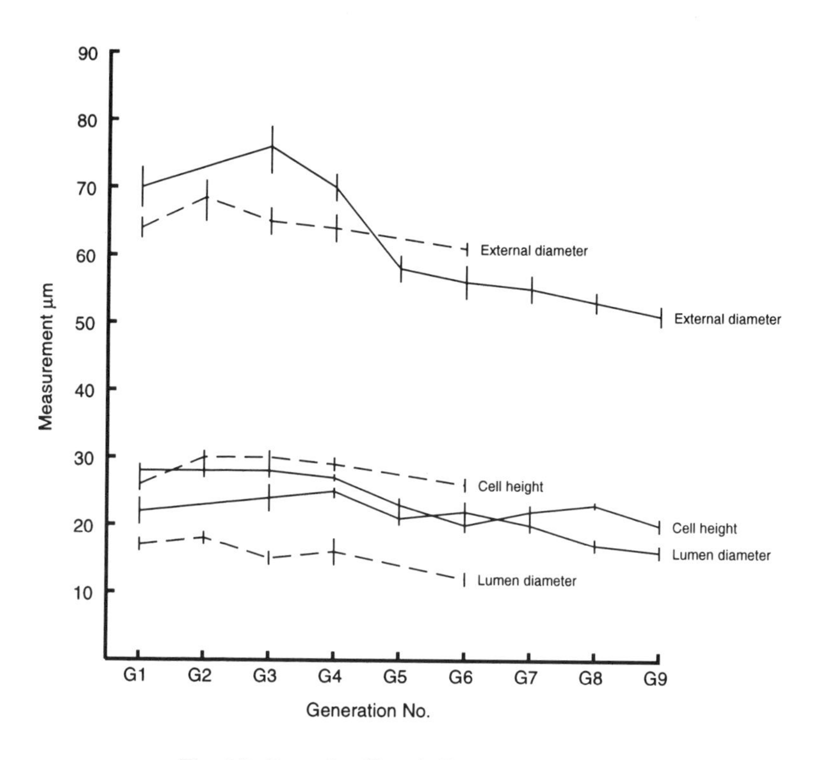

Fig. 6.5. Example of bars indicating means ± SE.

"mean plus or minus SD" (fig. 6.3b) or from the use of bars on graphs (fig. 6.5), but the reader cannot know whether ±2.5 is the SD or the SE. As both of these are derived from square roots, neither has a + or – sign anyway. 26mg (SD 2.5) or 26mg (SE 2.5) is a clear way to express it.

Statistical analysis

The raw and manipulated, descriptive or summarised data given above are sufficient for statistical analysis. It is not within the scope of this book to give a critique of statistical analysis. It *is* within its scope to give the researcher some simple landmarks and references with which to tackle the important phase of statistical analysis of his results. Coggon (1995) comments: "The good news is that...it is not necessary...to understand all the mathematical intricacies of a statistical calculation in order to apply its results"; but Gardner and Altman (1989) add: "Although nearly all medical researchers have some acquaintance with basic statistics, there is no easy way for them to acquire insight into important statistical concepts and principles.There is also little help available about how to design, analyse, and write up a whole project. Partly for these reasons much that is published is statistically poor or even wrong (Gardner *et al.*, 1983)". So you need statistical help, in the shape of "literature", computer programmes or people.

Statistical literature

There is plenty of literature, some of it designed specifically for statistically-challenged biomedical researchers, from Bradford Hill (1991), through Coggon (1995) and Hutchinson (1995) to Swinscow (1996). Bradford Hill (first edition 1937) and Swinscow (first edition 1976), in particular, went to great trouble to avoid the mathematical symbols that are the *lingua franca* of the statistician but which, I suspect, are the main reason (rather than our inherent innumeracy) that many statistical books are inaccessible to many of us. Perhaps this explains why Bradford Hill and Swinscow (both unfortunately now dead) are still in print. Swinscow specifically gives examples that can be handled on a pocket calculator. In fact this is a very constructive way of coming to grips with what one is actually doing in statistics (Montironi *et al.*, 1996).

Statistics by computer

There is also literature on how to get the computer to do the statistics for you, particularly Brown and Beck (1990) who show how to do it with two well known packages, Minitab and Statgraphics. Many other statistical packages are available for use on computers, for example big comprehensive ones such as SPSS (Statistical Package for the Social Sciences), and small ones for specific purposes such as Gardner *et al.*'s Confidence Interval Analysis (CIA)(1989) and KMSURV for survival curves (Prudente, 1988). The important point is to try to understand what the computer is doing statistically so as not to make statistical mistakes. As this may involve understanding both the statistical procedure and the computer procedure, it can be difficult.

Statisticians

Unfortunately, many biomedical workers do not understand statistics, and statisticians seldom understand biology, so that two-way communication is often difficult. It is not easy to get help from either amateur or professional statisticians. Amateur statisticians may know little more about how to handle your material than you do, and may lead you into error through lack of "insight into important statistical concepts and principles". If you seek help before you start from a professional statistician, who is likely to be busy, inaccessible, impatient and unfamiliar with your field, he may be able to help, particularly with the design of the project and how many cases you need to obtain the answer you require, especially if you have some results from a pilot study. Having gone so far, he may be able to advise you on how to analyse the results and what computer package to use. If you are especially lucky he may let you use his. If you seek help from a professional statistician *after* you have the results, he will almost certainly be critical of your design and methods; he may or may not be prepared to help you produce something publishable from the hopelessly inadequate results you show him. He may be able to advise on the most effective way of improving your data – for example, in a particular histological study, would it be better to look at more fields on your slides, cut more sections, or obtain more patients? (Analysis of variance showed the last to be the most efficient – Caroline Doré, personal communication).

Your aim must be to find a sympathetic co-professional, make friends with him, treat him well in terms of listening to and learning from him, and reward him with claret, praise and co-authorships, hopefully over many projects.

Types of statistical test

The data we collect are usually from samples of wider populations, and differences between samples can and do occur by chance. Chance can be quantified using the concept of probability (P). This is easily illustrated by thinking of tossing a coin which must come down heads or tails. With 100 tosses, 50 are likely to be heads and 50 tails, and we say that the chance of a head or a tail in any toss is 50:50, which can also be expressed as a probability of 50% or 0.5. A probability of 0% or 0.0 means never and 100% or 1.0 means always. Biomedical researchers may wish to determine the chances that a lump in the breast is malignant, or that a patient will survive breast cancer for five years, or that treatment with Tamoxifen will increase survival.

This concept of probability, which cannot be applied to an individual case (since who knows whether that case will come within the 90% of lumps that are benign or the 10% that are malignant), is refined for biomedical use by considering that diagnoses, for example, are themselves not wholly reliable and may turn out to be falsely positive or falsely negative. The percentage of cases correctly diagnosed by a clinical or laboratory test is called the "sensitivity" of the test and represents the probability that it will give a correct diagnosis. The percentage of cases in which the test correctly diagnoses that the subject does

Table 6.2. Coggon's (1995) example showing the use of sensitivity, specificity and predictive value.

Evaluation of mammography in the diagnosis of breast cancer

Outcomes in a series of 1500 women investigated by mammography

Result of mammography	Histologically proven tumour within six months of mammography		Total
	Yes	No	
Positive	9	59	68
Negative	7	1425	1432
Total	16	1484	1500

The sensitivity of mammography is the probability that it will correctly diagnose a true case, and is given by 9/16 = 56%

The specificity of mammography is the probability that it will correctly classify a non-case, and is given by 1425/1484 = 96%

The predictive value of mammography is the probability that a woman with a positive result really has a tumour, and is given by 9/68 = 13%

not have the disease is the "specificity" of the test and represents the probability that the subject does not have the disease. The "predictive value" of the test depends on the prevalence of the disease in the test population. Thus Coggon (1995) gives the realistic example (see table 6.2) of mammography in 1,500 women showing a sensitivity of 56% and a specificity of 96%, with the predictive value, the probability that someone with a positive test actually has breast cancer, as 13%.

Table 6.3 shows how the prevalence can affect the predictive value. In this case, reducing the prevalence from 1/1000 to 1/10,000 reduces the positive predictive value of whatever test was used from 9.9% to less than 1%.

Comparing assessments

We may wish to see whether there is inter-observer agreement in a test, for example among radiologists reading mammograms, as illustrated by Altman (1991) in table 6.4, and explained as follows. There were 54 exact agreements (21+17+15+1=54). The expected frequency in a cell of a frequency table

Table 6.3. Predictive value is affected by prevalence. (Courtesy of Professor Noah)

Sensitivity 99%
Specificity 99%

Prevalence 1/1000 (0.1%)

		Disease					
		positive		negative		total	PPV = a/a+b
Test	positive	a	990	b	10000	10990	9.90%
							NPV = c/c+d
	negative	c	10	d	990000	990010	<100%
			1000		1000000		

Prevalence 1/10000 (0.01%)

		Disease					
		positive		negative		total	PPV = a/a+b
Test	positive	a	99	b	10000	10090	<1%
							NPV = c/c+d
	negative	c	1	d	990000	990000	<<100%
			100		10000000		

Table 6.4. Inter-observer agreement.

	Radiologist B				
Radiologist A	Normal	Benign	Suspected cancer	Cancer	Total
Normal	21	12	0	0	33
Benign	4	17	1	0	22
Suspected cancer	3	9	15	2	29
Cancer	0	0	0	1	1
Total	28	38	16	3	85

Source: Boyd NF et al. (1982) J.Nat. Cancer Inst. ***68****, 357–63*

(under the null hypothesis of no association) is the total of the relevant column multiplied by the total of the relevant row divided by the grand total, i.e.:

normal	33 x 28/85 = 10.87
benign disease	22 x 38/85 = 9.84
suspected cancer	29 x 16/85 = 5.46
cancer	1 x 3/85 = 0.04
total	26.20

The number of agreements by chance is 26.2, which is 26.2/85 = 0.31. The possible scope for doing better than chance is 1.00 – 0.31, so agreement can be calculated as:

$$\frac{0.64 - 0.31}{1.00 - 0.31} = 0.47$$

This is known as the kappa statistic, which may be interpreted:

Value of Kappa	Strength of agreement
<0.20	poor
0.21–0.40	fair
0.41–0.60	moderate
0.61–0.80	good
0.81–1.00	very good

Statistical inference

Statistical inference is the process of dealing with the uncertainty in extending the differences in findings between sample populations to what might be happening in the population from which the samples have been taken, and including the possible effects of chance on those differences. This is done either by hypothesis testing or by estimation with confidence intervals.

Hypothesis testing

In the above tossing of a coin 100 times, our coin was only one sample of a whole population of coins that we might have used. Similarly, our sample of patients with boils was only one sample of a whole population of patients with boils, about all of whom we cannot possibly know. We start therefore by assuming that there is no difference between those given antibiotic or placebo: that is, we make a null hypothesis. We then define our level of statistical significance as the probability of obtaining a difference as large or larger than would have been observed if the null hypothesis were true. This is expressed as a P value, as shown below.

One or two group comparisons: t-test

To assess what the difference in results from two groups, such as our boil recovery time groups, means, we can test the *probability* that the difference occurred by chance. The first step is to determine whether the data in each group are normally distributed, since this will determine which test is used. This can be done by the graphical technique of normal plotting (Brown and Beck, 1990: pp. 18–19), which is laborious by hand but easy on a computer. If the data are not normally distributed they may become normally distributed by logarithmic transformation and thus suitable for the tests for normally distributed data.

If our two sets of data are normally distributed, we can apply the Student's t-test[1] for two unpaired groups, using the means and the standard deviations and the number of cases for each group. (A slightly different procedure is used if the data are *paired*, i.e. two successive tests on the same patient or specimen). The calculation produces a t value, with which from statistical tables in the statistic books (e.g. Swinscow, 1996) the P (probability) statistic (between 0 and 1) is obtained. The convention is that $P<0.05$, i.e. a less than 5% or 1:20 probability of the difference having happened by chance is regarded as significant. In our case there is a less than 1% probability that the antibiotic shortened the time by 4.4 hours by chance. If our two sets of data were not normally distributed, the Mann–Whitney U test or the Wilcoxon test are distribution-free tests for unpaired data and paired data. Different tests are used for nominal, ordinal and quantitative data. Seeking the advice of a statistician is usually sensible when choosing the correct test for your data.

The test can also be one or two tailed. In a one tailed test, we look at the deviation from our null hypothesis in one direction only, i.e. that the boil may get better more quickly in the antibiotic group. In a two tailed test, both directions are considered, i.e. that the boil may get better more quickly or more slowly in the antibiotic group. If not stated, a two tailed test is usually implied.

The problem with this approach is that it still considers only the statistics obtained from the samples of cases used. It does not consider estimates of the result that would be obtained if the total population, i.e. everyone who had a boil, had been included in the trial.

The confidence interval

This problem of considering the whole population of patients with boils, rather than just a sample, is, however, taken into account if we use the concept of the "confidence interval" (Gardner and Altman, 1989). This is a range of values within which we are fairly confident that the true value for the whole population lies. The range of the confidence interval depends partly on the standard error (the standard deviation divided by the square root of the number in the sample) and partly on the degree of confidence that we choose; just as we chose

1 Published in 1908 by WS Gosset under the pseudonym "Student" (Swinscow, 1996).

P<0.05 as our cut-off point for probability above. It takes the following form:

> the sample mean ± "some number" x standard error of the best estimate, i.e. the mean
>
> (for a confidence interval of 90%, "some number" is 1.64,
> for a confidence interval of 95%, "some number" is 1.96,
> for a confidence interval of 99%, "some number" is 2.58)

Thus the true number of hours by which the antibiotic shortened recovery lies between:

> 1.83 to 6.97 hours (confidence interval of 90%)
> 1.32 to 7.48 hours (confidence interval of 95%)
> 0.27 to 8.53 hours (confidence interval of 99%)

In table 6.1 the means, SDs, SEs and confidence intervals (90%, 95%, 99%) were obtained merely by typing the two sets of results into the CIA programme (Gardner *et al.*, 1989). Although we know the difference to be statistically significant, the confidence intervals allow clinical judgement to estimate the biomedical significance. The confidence interval is narrowed by having larger groups of cases, as shown in table 6.1 with 100 or 1,000 cases in each group.

Power

There are two ways of demonstrating the power of the study. First, the boil study showed reduction in healing time with the antibiotic:

Cases/group	95% Confidence interval
20	7.48 – 1.32 hours (6.16 hours interval)
100	5.74 – 3.06 hours (2.68 hours interval)
1000	4.82 – 3.98 hours (0.84 hours interval)

The larger the study, the smaller the interval, the more biomedically assessable the results and the greater the power of the study, but in retrospect.

Second, when planning the study the hypothesis testing approach is helpful. We saw in Chapter 1 that the null hypothesis might be rejected when it is true (type I error), or might be accepted when it is untrue (type II error). Furthermore, if the criterion chosen for rejecting the null hypothesis is lowered, from, say, P=0.05 (5%) to 0.01 (1%), the chance of type I error decreases, but the chance of type II error increases. The chances of type I and II errors can be reduced if the number of cases in the sample is increased. One (or

100%) minus the probability of type II error can be used as a measure of power. Thus the researcher may be able to say that with P=0.05 the study will have 80% power to detect an effect of a given size, but with P=0.01 it will have a 99% power to detect that effect.

Comparison of more than two groups

One may be tempted to do multiple t-tests on several groups of cases, for example comparing group A with group B, then with group C, then group B with group C. Unfortunately this ignores the associations inherent in multiple comparisons, and allows statistically significant differences to appear erroneously. This is avoided if the differences are tested by the *analysis of variance*. This is nicely explained in Brown and Beck (1990), with printouts from Statgraphics and Minitab packages, showing confidence intervals. The Kruskal–Wallis test, which offers a distribution-free method for testing multiple groups that are not normally distributed, is also discussed.

Are variables related?

Swinscow (1996) clearly explains how to test how closely two sets of variables are related. Altman (1991) shows how easy it is with the computer to draw a graph or "scatter plot" (fig. 6.6) with one set of variables along the x axis and the other along the y axis, and then to draw a regression line which shows the slope of the relationship. The equation of the regression line expresses the relationship between x and y (in the form of $y = a + bx$, where a = the point of interception of the y axis and b = the regression coefficient). The computer also calculates "r", the correlation coefficient, which lies between -1 and +1 (+1 or -1 indicate perfect correlation, while 0 indicates absolutely no correlation), the P value, and draws the lines alongside which depict the confidence interval at any point along the regression line.

Life tables and survival curves

Figure 6.7 shows a common way of relating any variable or variables to prognosis in the form of survival curves which are derived from life tables. A method of comparing survival curves to assess whether they are statistically significantly different is also needed. This can be done using the logrank test (Mantel, 1966). The theoretical basis of these descriptive tables or graphs of observed outcome, and how to construct and test them is not easy to under-

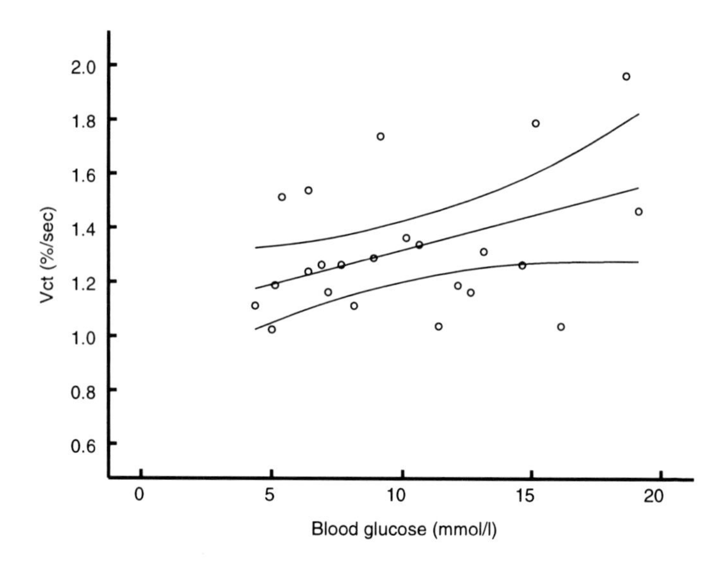

Fig. 6.6. "Scatter plot" with regression line and confidence intervals.

stand. It is, however, possible to follow the long paper of Peto *et al.* (1977) and gain some insight into both. An important and practical point is that all the patients in a study can be entered into the analysis, whether they have died or reached some other designated end point or not, provided that whether or not they have reached it and the duration of their time from the designated start point are known.

Many of the big statistics packages have a survival curve component, but KMSURV, a shareware program which can be obtained for the price of the postage (US$5),[2] is an easy way into this form of analysis. Simply entering the case, the group it is in, whether it has reached the end point and its duration from the start point is enough for the programme to draw the Kaplan–Meier survival curves and print out the statistical significance between them.

This is encouraging, since much work is done on, for example, cell variables in cancer, which, to be of any value, must be related to the behaviour of that cancer, in terms of killing the patient or metastasising, and to the collection of information about what happens to the patient, in which effort must also be invested. The CIA programme of Gardner *et al.* (1989) can also be used to give confidence intervals for survival time analysis.

2 KMSURV, from Epidemiology Monitor, 2560 Whisper Wind Court, Roswell, GA 30076, USA

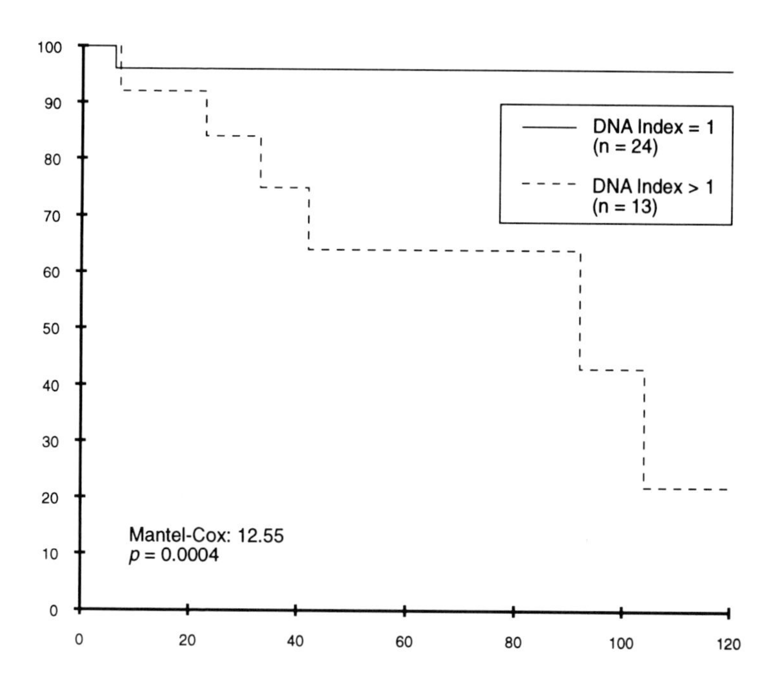

Fig. 6.7. Comparison of survival curves.

If the relative effects of several variables are to be combined to decide on the treatment of the patient, the various techniques of multivariate analysis can be applied. For example, in breast cancer, the diameter of the tumour, whether lymph node metastases are present or not, and the mitotic activity index (MAI) of the cancer cell nuclei can be combined mathematically on the basis of a series of cases. In one study (Baak *et al.*, 1982), a multivariate prognostic index (MPI) was produced as follows:

MPI = +0.3341 x MAI
+0.2342 x tumour diameter (cm)
–0.7654 x lymph node status (positive = 1; negative = 2)

For a new patient, if the MPI was less than 0.6, the probability of 10-year survival was 88%; if equal to or more than 0.6 it was 46%.

Bayesian statistical methods

A further approach to extracting better decision making from the statistical data is through Bayesian analysis, which is able to incorporate additional data as it becomes available. The basis for this approach is set out by Lilford and Braunholtz (1996), and another worked example is shown by Montironi *et al.* (1996).

Develop your statistical independence

A great deal more can be done with statistics, but you must remain in touch with your readers, who must be able to follow your statistical reasoning, since upon it will depend whether they can believe that your data confirm the truth of your various messages. As many of the concepts are not easy to grasp, it is helpful to have access to a variety of explanations. I have found something of help in the references below, and in the statistics books listed under Further Reading on p. 171.

References

Altman D. Practical statistics for medical research. London: Chapman and Hall, 1991.

Baak JPA, Kurver PHJ, Snoo-Nieuwlaat AJE. Prognostic indicators in breast cancer: morphometric methods. Histopathology 1982;6:327–39.

Bradford Hill A, Hill ID. A short textbook of medical statistics. 12th edn. London: Hodder and Stoughton, 1991. (First published as "Principles of medical statistics", London: The Lancet, 1937).

Brown RA, Beck JS. Medical Statistics on microcomputers. London: British Medical Journal, 1990.

Coggon D. Statistics in clinical practice. London: British Medical Journal, 1995.

Gardner MJ, Altman DG. Statistics with confidence. London: British Medical Journal, 1989.

Gardner MJ, Altman DG, Jones DR, Machin D. Is the statistical assessment of papers submitted to the British Medical Journal effective? British Medical Journal 1983;286:1485–8.

Gardner MJ, Gardner SB, Winter PD. Confidence interval analysis (CIA). London: British Medical Journal, 1989 (with 5.25-inch or 3.5-inch computer disk, available from the British Medical Journal, Tavistock Square, London WC1H 9JR, UK).

Hutchinson TP. Version 4 (Health and Sickness) of essentials of statistical

methods. Sydney: Rumsby Scientific Publishing, 1995.
Kirkwood BR. Essentials of Medical Statistics. Oxford: Blackwell, 1988.
Lilford RJ, Braunholtz D. The statistical basis of public policy: a paradigm shift is overdue. British Medical Journal 1996;313:603–7.
Lock S, Wells F. Fraud and misconduct in medical research. London: British Medical Journal, 1993.
Mantel N. Evaluation of survival data and two new rank order statistics arising in its consideration. Cancer Chemotherapy Reports 1966;50:163–70.
Montironi R, Whimster WF, Collan Y, Hamilton PW, Thompson D, Bartels P. How to develop and use a Bayesian Belief Network. Journal of Clinical Pathology 1996;49:194–201.
Peto R, Pike MC, Armitage P, Breslow NE, Cox DR, Howard SV, Mantel N, McPherson K, Peto J, Smith PG. Design and analysis of randomized clinical trials requiring prolonged observation of each patient: II Analysis and examples. British Journal of Cancer 1977;35:1–39.
Prudente A. KMSURV: a microcomputer program for univariate survival data analysis. Sao Paulo: Ludwig Institute for Cancer Research, 1988.
Smith R. Steaming up windows and refereeing medical papers. British Medical Journal 1982;285:1259–61.
Swinscow TDV. Statistics at Square One. 9th edn. London: British Medical Journal, 1996.

7. The visual display of data

Excellence in statistical graphics consists of complex ideas communicated with clarity, precision and efficiency.

Edward Tufte (1983)

Research, scholarship and indeed thought are very closely bound up with seeing and visualising. The other four senses, touch, hearing, smell and taste, play much smaller parts, even in medical practice. I often ask classes of undergraduates if any of them do their thinking or imagining other than in images. They all think in visual images, mostly in colour. Many can also imagine music or sounds, smells, tastes and touches, but do not think in those modes. To me the months in the calendar and years in past and future decades are shaded light and dark, but not coloured. I also run through memories in my "skull cinema", as described by Hillaby (1970). I cannot, however, run through musical scores in my head, as Vaughan Williams did on long train journeys.

Imaging techniques play an increasing part in biomedical work. Laboratory results are displayed visually as columns of figures, which may be converted automatically by the computer into one, two, or three dimensional forms to aid interpretation and understanding of relationships. This is hard visual display material, in that it is presented similarly to every viewer, whatever he may make of it.

Writing is soft visual material from which the reader has to create his own mental displays; these cannot be monitored or compared. The feeling is analogous to doing microscopy with one's supervisor in the days before double-headed microscopes and pointers. *Are we talking about the same cell? How do I select the significant ones from the multitude in the field?* Nevertheless, writing can convey a very visual message, for example "A woman with the stiff man syndrome" (Asher, 1958). It can also convey a vivid but inaccurate message, for example "Gold lung", which was about a case of reversible pseudo-malignancy in the lung caused by aurothiomalate therapy (James *et al.*, 1978), not a chest physician with the Midas touch.

Representation of mental displays is the field of the artist, processing selectively what he has seen or read or imagined into another hard visual display,

Fig. 7.1. Medical art from the days before photography.

pressed extensively into the service of science in the days before photography, for example Wright of Derby's 'An experiment on a bird in an air pump' (1768) (fig. 7.1). Medical artists in recent times, for example Simmonds and Reynolds (1983, 1989, 1994), have progressively pressed modern technology back into a do-it-yourself service for researchers.

Properly understood, therefore, visual display of your data is a very powerful tool, but you have to think about how to use it, because it is often used thoughtlessly and ineffectively. Visual display is not usually capable of conveying the message in a publication. It is used to convey the data with which you hope to convince the reader that the message is to be believed and even acted upon, whether in a journal article, an oral presentation or a poster. This is the usual purpose of visual display, and amid the mass of data and the wealth of modern display techniques, this purpose should be kept firmly in mind.

> Data graphics visually display measured quantities by means of the combined use of points, lines, a coordinate system, numbers, symbols, words, shading and color... The use of abstract, non-representational pictures to show numbers is a surprisingly recent invention...not until 1750–1800... Modern data graphics can do much more than simply substitute for small statistical tables. At their best, graphics are instruments for reasoning about quantitative information. Often the most effective way to describe, explore, and summarise a set of numbers is to look at pictures of those numbers.
> (Tufte, 1983)

Such visual display is required by the biomedical scientist in the three standard circumstances of this book, namely, for papers for publication, for oral presentations and for posters. He often wishes to present his data in all three forms. The same principles apply to each, but there are different constraints. The author needs to understand these principles and constraints because he is not only expected to produce the displays himself in a camera-ready form, i.e. ready for printing without further modification, but in many cases has access to the largely computerised tools to do it easily. Indeed, as Simmonds reported (1983), provision of the materials and equipment by the medical illustration department in his institution allowed 1,400 authors to produce 6,000 illustrations for over 1,400 papers and presentations in a year, ten times the previous output of the professional audio-visual staff. Although referees may criticise the displays, it is now very unusual for journals to be prepared to redraw them for the author, whereas the journal office does decide the details of the print typeface and size and layout for the presentation of the text, so that it is consistent throughout the journal. The description of typefaces and so on below can be applied to text but is intended for you to apply to your visual displays.

Visual display in journal papers

Here the main constraint is page space. Tables are used for raw and manipulated data, and relational graphics for relating data sets to each other to show relationships, possibly causal ones. You will probably need both.

Tables

Tables (fig. 7.2) consist of rows read from left to right and columns read from top to bottom. Readers find it convenient to have the identification information for the rows sited down the left side, and for the columns across the top. In the computer, a table can easily be constructed using the spreadsheet format, with rows and columns of boxes into which text or numbers can be typed. These

Horizontal and vertical emphasis

	1	2	3	4	5
A	28.5	38.5	48.8	12.5	4.5
B	43.9	43.9	43.0	8.9	43.9
C	16.0	116.0	46.0	45.0	5.0
D	8.6	18.6	78.6	8.6	8.6
E	14.6	24.6	34.6	14.0	14.6
F	39.0	39.0	9.0	69.0	39.0
G	105.9	45.9	135.0	200.0	45.9
H	9.9	10.9	49.0	8.9	9.9
I	23.0	65.0	46.0	67.0	6.0

Too much space between the columns makes it difficult to read across the rows

	1	2	3	4	5
A	28.5	38.5	48.8	12.5	4.5
B	43.9	43.9	43.0	8.9	43.9
C	16.0	116.0	46.0	45.0	5.0
D	8.6	18.6	78.6	8.6	8.6
E	14.6	24.6	34.6	14.0	14.6
F	39.0	39.0	9.0	69.0	39.0
G	105.9	45.9	135.0	200.0	45.9
H	9.9	10.9	49.0	8.9	9.8
I	23.0	65.0	46.0	67.0	6.0

Vertical rules also make it difficult to read across the rows

	1	2	3	4	5
A	28.5	38.5	48.8	12.5	4.5
B	43.9	43.9	43.0	8.9	43.9
C	16.0	116.0	46.0	45.0	5.0
D	8.6	18.6	78.6	8.6	8.6
E	14.6	24.6	34.6	14.0	14.6
F	39.0	39.0	9.0	69.0	39.0
G	105.9	45.9	135.0	200.0	45.9
H	9.9	10.9	49.0	8.9	9.9
I	23.0	65.0	46.0	67.0	6.0

More space between rows, possibly with thin horizontal rules, will help with reading across. However, inserting rules may conflict with the house style of journals. Check with the editor first.

Fig. 7.2. Horizontal and vertical emphasis in a table of data.

can be ranged left, right or centred. The actual boxes are not usually printed out. If you are without computers or spreadsheets, it is best to range everything left because inaccurate manual centring is definitely distracting. In any case the reader's eye is trained to start at top left. There are various graphical devices for grouping related data or separating groups of data, but the rule is only to include extra row or column lines if they have a clear function (fig. 7.2). As Tufte (1983) puts it, erase "non-data ink" if you can.

The author does, of course, have to decide on the type size and typeface for camera-ready (ready for printing without further typesetting, but may be reduced in size) tables and graphs. Most computer word-processing packages offer a range of type sizes (mine offers 14, from 6 point to 40 point, see Appendix 5) with easy conversion from one to another on the screen. The unit of measurement "point" (pt) is a historic size; there are about 72 points to the inch (25mm), so one point is about 0.35mm, and 10pt is about 3.5mm. The print size is the height of a capital letter. In former times each letter had to be cast in metal separately in a mould and mounted in rows in a frame for printing. The space between the lines was created by the type body which extended around the letter face. So "10 on 11pt" would mean a 10pt letter cast on an 11pt body. In word-processing packages you can make whatever space you like between lines, by setting the "leading", "line spacing", "inter-line spacing" or "linefeed". With, say, 10pt type, 10pt linefeed will just separate the letters, so 11pt linefeed might be more legible.

My word-processing package also offers a number of different typefaces or fonts (Appendix 5), again with easy conversion of text from one to another, so the author can easily try them all out. Four are "serif" typefaces (with finishing shapes at the ends of letters) and two "sans serif" (without finishing shapes). Serifs are supposed to give horizontal emphasis to guide the eye along the line. The thin finishing shapes may disappear with photocopying or photographing. The characters in most typefaces have a *variable* width. There are a few typefaces where all characters have a fixed width, so that an "m" looks constricted and an "i" has too much space (see Courier, Appendix 5). Sans serif typefaces have a business-like appearance and are good for photography or photocopying, unless the strokes are thin as in Courier.

The word-processing package can also reproduce each normal typeface in different styles, particularly **bold** and *italic* or ***both***, which can be used judiciously to provide emphasis. Bold type photocopies and photographs well. The package can also underline but this reduces legibility by cluttering the space between the lines. Some packages have shadow and outline styles, but these give decoration rather than content and reduce legibility.

On the whole readers do not like mixtures of typefaces. They can, however, take the same typeface in different sizes for specific purposes, such as hierarchies of headings, and some use of the bold or italic styles, provided it is purposeful and not used idiosyncratically, as some people use exclamation marks!

Legibility is difficult to define or investigate, but rate of reading has been found to be the best measure. It is affected by type size. For ordinary hand-held reading, 8pt is too small, 14pt is rather big. This page is set in 10pt. The typeface or font does not affect legibility much. Readers are more affected by familiarity and aesthetic preference. Printing text all in capitals reduces the reading rate considerably. So mixed capitals and lower case letters help legibility, as does black type on an unpatterned white background.

In a table you are constrained by the number of items of data you need to present. You may be able to get more items in if you use a smaller typeface, which will affect legibility, but you should also remember that the journal may reduce the size of camera-ready copy to perhaps 75% of the original, and that this will also reduce the thickness of any ruled lines. This may affect your choice of a typeface with narrow lines such as Courier. It is very important to study the Instructions to Authors of the target journal for instructions on typefaces or sizes, and also to look, critically, at any tables that you can find in the text in that journal.

Table design

You must decide what you are trying to convey to the reader. How many items of information do you want to show? You can show a lot but do you really need to show all of them? Should you use "portrait" orientation, i.e. with the long axis vertical, or "landscape" orientation, i.e. with the long axis horizontal? What about the title and caption, if any? What about explanatory notes? Readers do not like to have to search the text for explanations.

You may use tables to record quite disparate items. For example fig. 7.3 shows the details of patients in numerical sequence by row and orientated in "portrait" mode, since there are 10 rows and 4 columns (40 items of informa-

Patients studied

Case	Sex/ Age	Smoker	Diagnosis
1	M/42	pipe	Oat cell carcinoma
2	M/62	15/day	Oat cell carcinoma
3	F/54	non	Aspergilloma
4	M/52	20/day	Squamous carcinoma
5	M/40	60/day	Adenocarcinoma
6	F/52	15/day	Squamous carcinoma
7	M/63	30/day	Adenocarcinoma
8	M/63	50/day	Squamous carcinoma
9	M/77	28/day	Squamous carcinoma
10	F/58	5/day	Adenocarcinoma

Fig. 7.3. Data oriented in portrait mode.

Distribution of distended, intermediate and flat cell acini						
	Total acini	Distended	Intermediate	Flat	χ^2	P
χ^2 test vs own control readings						
(n = 12)						
Control	1351 (100%)	448 (33.2%)	858 (63.5%)	45 (3.3%)	25.09	<0.001
DBcAMP	1116 (100%)	433 (38.8%)	612 (54.8%)	71 (6.4%)		
(n = 4)						
Control	675	169	256	21	22.47	<0.001
Theophylline	513 (100%)	125 (24.4%)	196 (67.6%)	17 (5.9%)		
(n = 4)						
Control	409	174	252	15	14.02	<0.001
Opipramol	290 (100%)	77 (26.6%)	196 (67.6%)	17 (5.9%)		

Fig 7.4. In this example, data is oriented in landscape format.

Cases and discharge indexes (DI) arranged in ascending order of gland wall ratio (GWR)

Case	Site	GWR	Cigs/day	DI Control	DI DBcAMP	DI Theophylline
5	RM	0.27	60	26.0	42.9	55.0*
3	RL	0.34	0	14.3	23.1	
10	LM	0.34	5	15.6	25.9	28.9
13	LL	0.40	0	11.6	25.8	
9	RU	0.41	28	10.8	9.6	21.6
17	RU	0.42	50	17.5	6.1	
2	LM	0.43	15	38.4	40.5	
1	LM	0.46	pipe	1.8	40.5*	
8	LU	0.46	50	22.9	31.9	33.3
4	RU	0.51	20	6.1	17.4*	

RM = right main bronchus
LM = left main bronchus

RU = right upper lobe bronchus
LU = left upper lobe bronchus

RL = right lower lobe bronchus
LL = left lower lobe bronchus

*significantly different from control; P<0.05

Fig. 7.5. A table arranged in a "reader-unfriendly" way.

tion). Figure 7.4 shows the raw data for a series of chi-squared tests plus manipulated data of the results of the tests and the P values obtained (55 items of information) in "landscape" mode. In both figs 7.3 and 7.4 it is good that the reader's eye is led along the rows to the answers to the tests.

In fig. 7.5, however, the reader may panic because the first thing he notices is that the cases (left-hand column) are not in sequence. What is the rationale? It is explained in the title, but the reader has to search for the vertical sequence under GWR (gland wall ratio). There is conflicting logic here. It is logical to go from the case to the site in the bronchial tree to the GWR, but not logical to have a non-sequential list of cases. It would not improve the table to have the GWR column on the left, but it would be logical to put the GWR column in bold. This is called "typographic cueing".

The principle is to do what is reader-friendly. If in doubt, select a helpful reader to tell you in what ways you could make the table more accessible.

Charts and graphs

The terms "charts" and "graphs" seem to be used interchangeably in these computer days. Whereas tables are suitable for storing numerical data for the readers' use, readers can be led to their meaning most conveniently via pictures, for example bar charts or bar graphs (fig. 7.6a and b), which compare different variables at one time, or column graphs, which compare the same variable at different times or in different groups. Column graphs can be used to show the means and standard deviations or standard errors, i.e. the manipulated data, but it has to be made clear that the horizontal top represents the mean value and whether the bar represents standard deviation or standard error (fig. 7.7). Histograms (fig. 7.8) also look like column graphs but, strictly speaking, they illustrate continuous data and compare the areas under the graph. This is not reader-friendly without actual numbers. Tufte (1983) interestingly points out that the bar of the bar chart proclaims its statistic no less than four times: by the heights of the left and right verticals, the horizontal top, and, possibly misleadingly if it is not a histogram, by the area under the horizontal. He suggests that such redundancy is unnecessary, but even a thin line is a form of bar and a bar of suitable thickness for reproduction is surely necessary. However, I agree that decoration of the bars with the various forms of shading available in the graphing packages (a form of what he calls "chartjunk") is usually not necessary (fig. 7.9) unless groups of bars need to be distinguished, for example within A, B, C, D, and E in fig. 7.9c. Pie charts are quite good for showing the distribution of items within a total (fig. 7.9d), but the human eye finds it difficult to compare distributions in two or more pie charts side-by-side or for too many or too small slices (fig. 7.10), and they are not enhanced by being three-dimensional. Three-dimensional charting is usu-

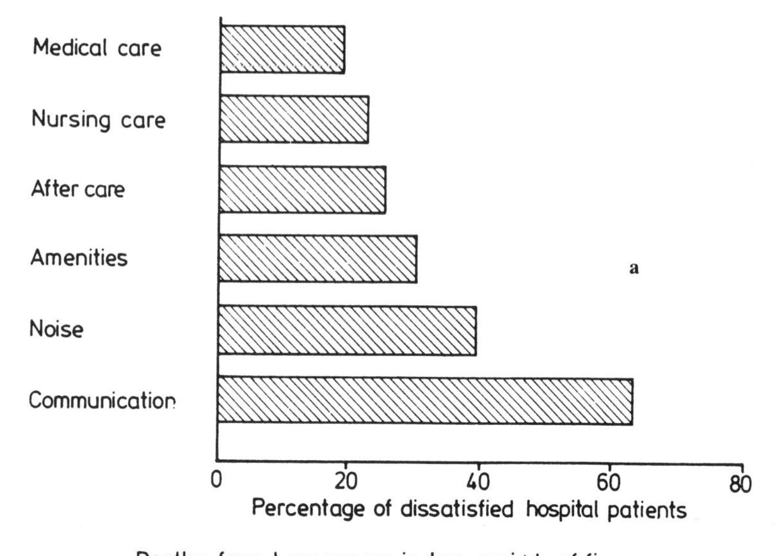

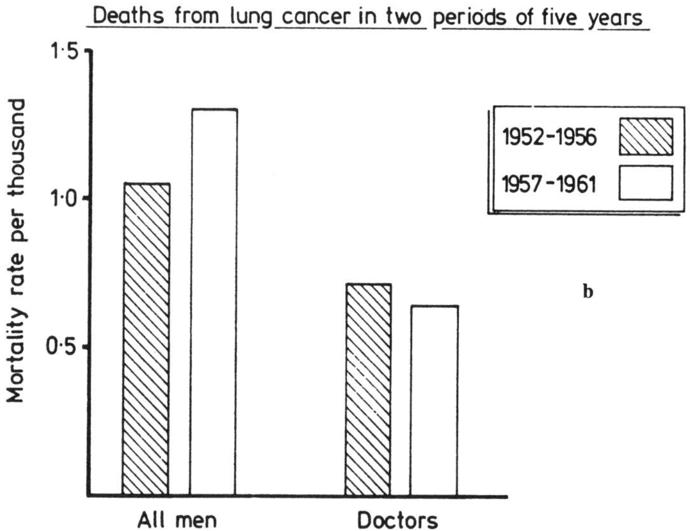

Fig. 7.6. Bar graphs (**a**) and column graphs (**b**) are convenient ways to illustrate data.

ally also unnecessary when two-dimensional variables are under consideration (fig. 7.9), but can be very useful for three dimensional data (fig. 7.11).

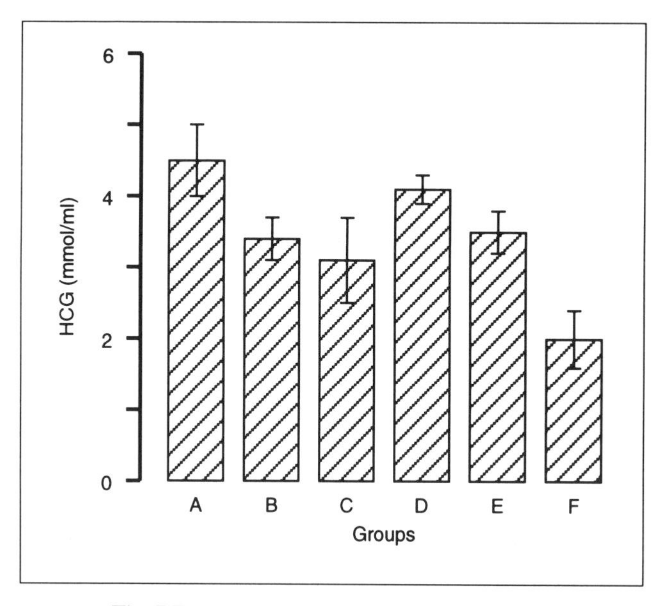

Fig. 7.7. Column graph showing standard errors.

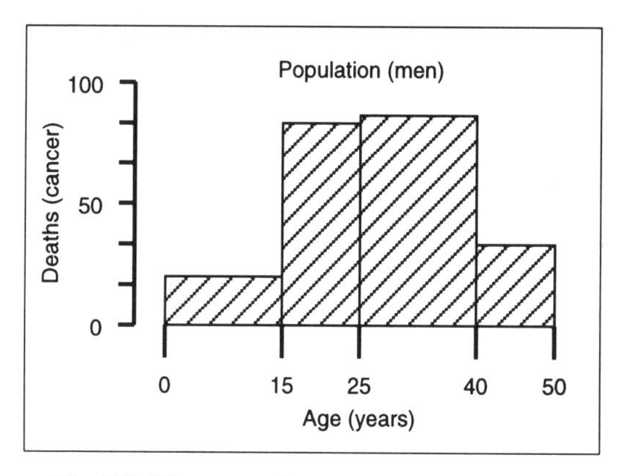

Fig. 7.8. Histograms illustrate continuous data and compare the areas under the graph.

The dot plots (vertical scatter plots) shown in fig 6.4 are a variant of the column graph, but give much more information, including two distributions,

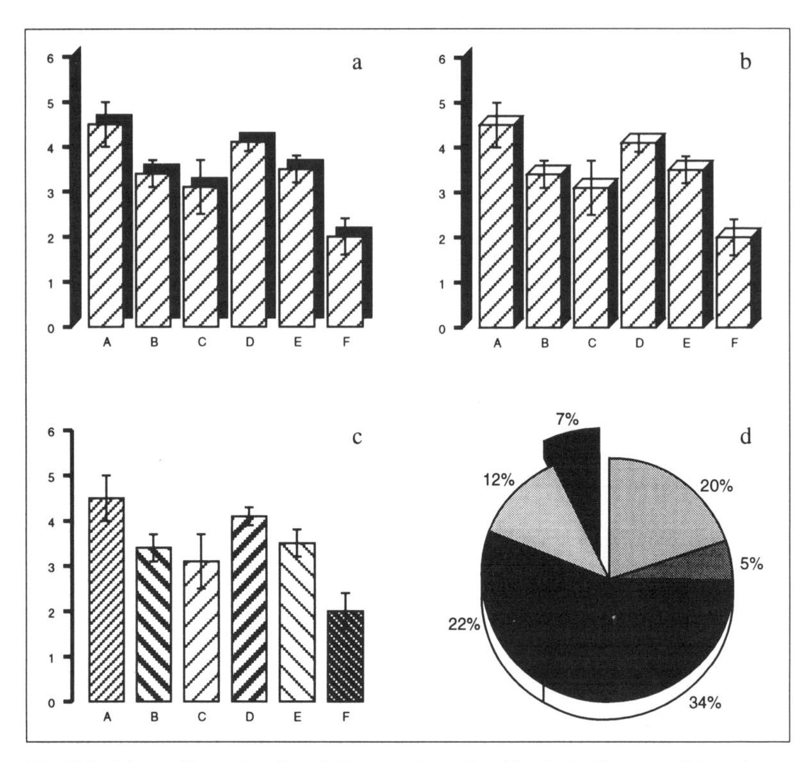

Fig. 7.9. Three-dimensional and decorated graphs. The 3-d effects used in column graphs **a** and **b** and pie chart **d** do not help to make the data more accessible. The shading used in column graph **c** can help to distinguish one item from another.

ranges and outliers, to which the arithmetic means, medians, modes and quartiles are easily added.

These charts can all be constructed manually, but computer graphing packages make it much easier. The so-called independent variable, often time, is by convention (i.e. what readers are used to) plotted along the horizontal (x) axis, and goes into the first column on the spreadsheet. The dependent variable, such as the haemoglobin level, goes along the vertical (y) axis. The tick marks that indicate the intervals on the x and y axes should face outwards to avoid any data on the graph nearby. Some graphing packages are linked with the data so that entering changes in the data automatically changes the graph.

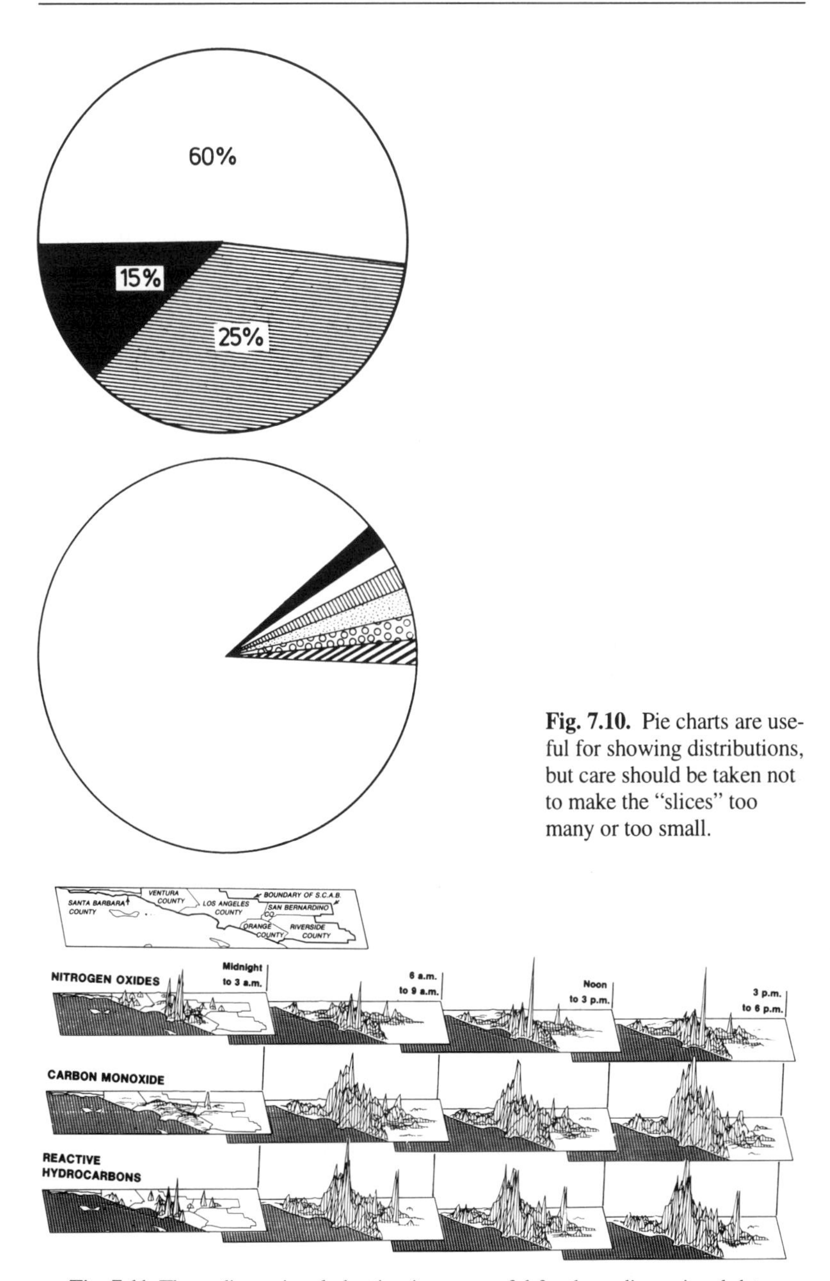

Fig. 7.10. Pie charts are useful for showing distributions, but care should be taken not to make the "slices" too many or too small.

Fig. 7.11. Three-dimensional charting is very useful for three-dimensional data.

Relational graphics

Visual graphics started 5,000 years ago with maps drawn on clay, which led to charts showing data concerning winds or currents in the 17th century. Dr John Snow used a map of central London to visualise the distribution of deaths from cholera in 1854. In 1975 maps showing the cancer mortality in 3,036 counties of the USA 1950–1969, each containing 21,000 items of data, were published (fig. 7.12).

Time series charts, for example relating soil temperature to depth over time, came next. But although Tufte says that descriptive chronology rarely leads to understanding of the causation, nevertheless it can be the way to show the complicated course of a disease in a patient (fig. 7.13).

Third, narrative graphics may be constructed to show changes in space and time, again with time along the x axis, as for example, Marey's 1885 map of Napoleon's invasion of Russia, showing the reduction over time and temperature of his army from 422,000 men at the Niemen river in Poland to 100,000 in Moscow and 10,000 on reaching the Niemen river again. This is graphical ingenuity of a high order.

Relational graphics, not tied to the physical world of geography or time or space, were developed in the 19th century and allow any variable quantity to be placed in relationship to any other variable quantity. The scatter plot (Figure 7.14) is the simplest form of relational graphic. It links at least two variables,

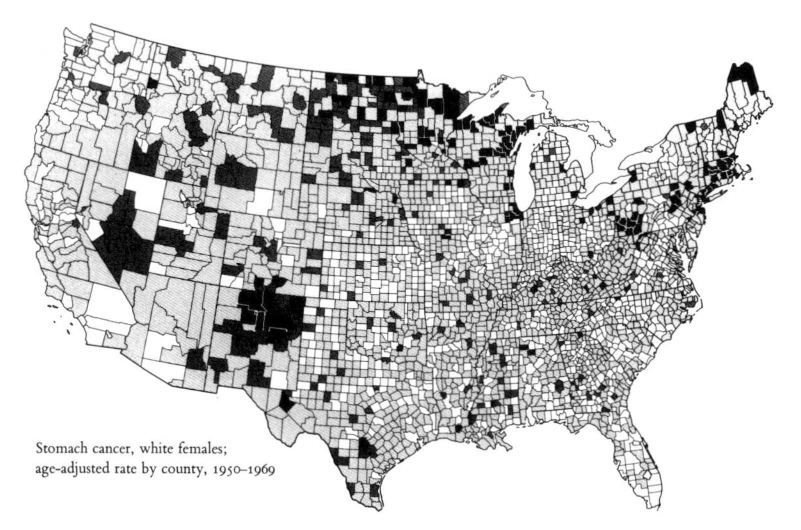

Fig. 7.12. Maps are a useful device to show, for example, the relation of items of data between one part of the country and another.

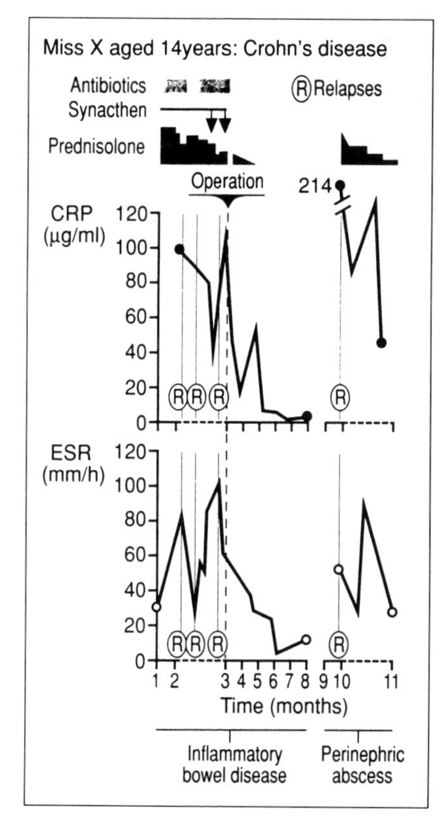

Fig. 7.13. Time series charts can plot the progression of disease in a patient.

"encouraging and even imploring the viewer to assess the possible causal relationship between the plotted variables", as shown in lung cancer and smoking (Tufte, 1983). Figure 7.14 also shows statistical treatment of the data (regression analysis) helpfully superimposed on the graphic.

Unfortunately, graphics depicting statistical data can mislead or even lie in five ways (although Benjamin Disraeli only allowed for three kinds of lies: "lies, damn lies and statistics"):

- representation of the numbers on the graphic not directly proportional to the numbers
- labelling on the graphic inadequate
- design variation rather than data variation
- the number of information variables exceeding the number of dimensions in the data

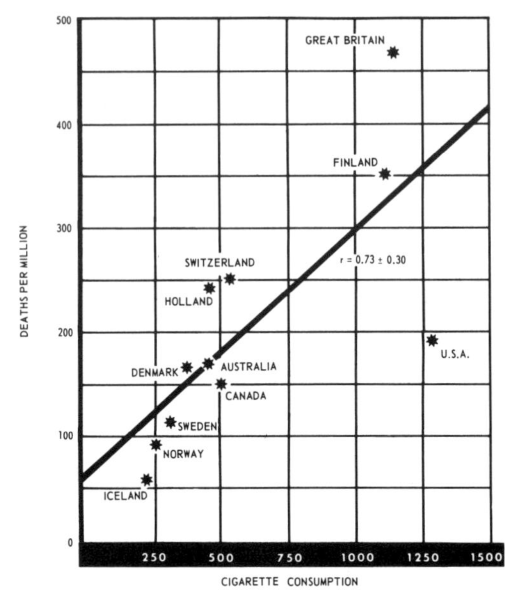

Fig. 7.14. The scatter plot is the simplest form of relational graphic.

- quoting data out of context

To be sure of reader-friendly, honest graphic design for your work, you should:

- look at all the tables to formulate a common design
- make sure that the numbers on the graphic are directly proportional to the data numbers
- ensure that the independent variable is on the x axis
- show a break in the y axis if it does not start at zero
- erase as much "non-data ink" as you can
- eliminate as much chartjunk as you can
- use no more than one typeface, but type size can be varied
- only use recognised abbreviations
- test it on a reader

Photographs

A clear photograph can often greatly help reader-understanding and be "worth a thousand words", but is it really essential to your message or its proof? If so, it must show what you want it to show and as little else as possible. This means making a print at a suitable magnification and cropping unnecessary surrounding material, especially if this brings it to the exact width of a column or columns in the target journal, which will avoid enlargement or reduction.

If a micrometer marker bar is needed to indicate scale, it should be put directly on to the print with Letraset so that it participates in any enlargement or reduction. Other identification marks, such as arrows or letters, are usually essential for the reader to be sure that he is looking at the right structure, but they should be put on to a tracing paper overlay for the printer to incorporate, not directly on to the print, as should marks to conceal a patient's identity or any other unwanted clinical detail. The author's name, the figure number and the orientation (by an arrow to the top) should be shown on the back of each print, not by writing directly on the print which may cause indentations visible on the front, but by affixing a previously prepared sticky label. Any textual matter, such as the caption or histological stain, should be typed in the figure legend list on a separate sheet. Copies of the sets of prints equivalent to the number of copies of the text required should be submitted to the journal, preferably in separate labelled envelopes without paper clips (they mark the prints) for easy dispatch to the referees. Photographs of x-ray films are particularly difficult to reproduce because they lack contrast and usually have to be drastically reduced; they are difficult to label, and a simple line drawing alongside may be the best solution.

Naturally, any advice available in the Instructions to Authors (Appendices 1 and 2) should be followed carefully and the quality of the photographs in previous numbers of the journal studied. Indeed for specialities that rely on images, such as histopathology and radiology, the quality of the photographs may well be one of the factors influencing the choice of target journal. The advice of an informed reader is well worth seeking to ensure that the photographs are of good enough quality, in terms of clarity, focus, contrast, magnification, size, identification of the components under discussion, and, for histopathology, clean white background. If colour is important for discriminating between or identifying components, the author may have to pay for the printing, but this may be worthwhile.

Visual displays for oral presentations

Visual displays in oral presentations include 35mm transparencies for direct projection and large transparent plastic sheets for projection on an overhead projector. To avoid confusion I call the 35mm transparancies *slides*, and the plastic sheets for overhead projection *OHPs*, although some people also call these slides. Overhead transparency projectors may also have a roll of transparent plastic sheet for speakers to draw or write on during the course of their talks. It is possible, but distracting for the audience, to use slides, OHPs and overhead transparency writing/drawing in the same presentation, but see Chapter 19 for advice about this.

OHPs offer the freedom of being easy to make yourself on your own premises at the last moment, using the departmental photocopier or a small, cheap, separate machine (mine cost £100 in 1976 and is still going strong). Some photocopiers can produce colour transparencies. Slides, on the other hand, need more forethought if you have to get other people to take and process the photographs.

Designing visual displays

You will find that everyone, and all "How to do it" books, have their own ideas about designing slides. You have to find out what works for you, since, as Simmonds and Reynolds (1994) say, "a slide presentation is a theatrical performance, a one-person show, with the audience wanting to be delighted with what they are seeing."

I start by designing my talk, working out my message(s) and what supporting evidence I need, and my conclusions, in list form on A4 ruled feint paper with plenty of lines. I then list all the visuals (slides and/or OHPs) I would like to have to illustrate my show. I do not worry about the number of visuals at this stage because I cannot tell how long I will spend on each one, although I know that audiences take in pictures much more quickly than words or graphs or tables, especially if I have an effective pointer. I know that double or triple projection uses many more slides and may become unsynchronised.

I think about my audience. I know that conference audiences like slides best because they seem to be more professional and imply more conscientious preparation (and hence respect for the audience). OHPs may be more difficult for large audiences to see, and changing them is distracting. Workshop audiences and small groups find OHPs more spontaneous, and you (and they) can even write on them as the workshop develops. Small groups can also cope with a combination of OHPs and slides, but you have to make the changeovers slick. I know that I can use slides to show words, graphics and pictures, but that OHPs, while showing words and graphics well, present pictures unattractively and usually without full colour.

I see what visuals I already have and what I will have to make – usually at least a new title one and a new conclusion one. I set out to design the new word and graphic slides, knowing that the proportions of the 35mm format are 3:2 (a

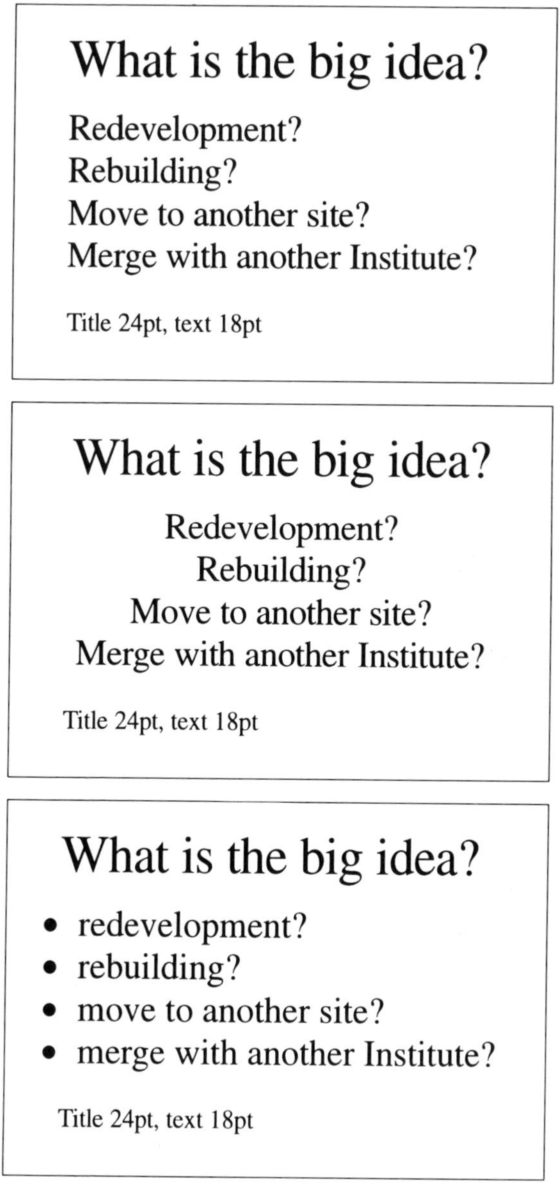

What is the big idea?

Redevelopment?
Rebuilding?
Move to another site?
Merge with another Institute?

Title 24pt, text 18pt

a

What is the big idea?

Redevelopment?
Rebuilding?
Move to another site?
Merge with another Institute?

Title 24pt, text 18pt

b

What is the big idea?

- redevelopment?
- rebuilding?
- move to another site?
- merge with another Institute?

Title 24pt, text 18pt

c

Fig. 7.15. Three possibilities for setting out a word slide.

horizontal A4 sheet, 297 x 210mm, 11.7 x 8.3in, has roughly the same proportions) and that slide printing devices need horizontal orientation. It is usually best to plan for horizontal projection as well, because in many lecture theatres vertically projected images extend irritatingly above and/or below the screen.

To make a word slide, create a 3:2 area on your computer screen (mine works out at 144 x 96mm). For positive slides, black on a clear background, 14pt text with 18pt titles will be legible, but while black on white is acceptable on OHPs, audiences have become used to colour on slides. So for negative slides, including diazos, with white or colour on a background of a different colour, 18pt text with 24pt for titles will be legible. I find white on blue diazos easy on the eye, although one study (Snowberg, 1973) found that a blue background gave the worst legibility. I find multiple colours confusing and the red end of the spectrum hard to see. Don't forget that up to 8% of men and 0.4% of women have some form of colour blindness, mostly inability to distinguish between red and green. Fancy backgrounds come into Tufte's category of chartjunk and should be avoided if you have worthwhile content and are not merely trying to entertain. Figures 7.15a, b and c show that the type size (using capitals at the left, and certainly not all capitals) and how it looks to you determine how much you can put in your space, so rules about how many words you should have on a slide can be ignored. You can decide whether to have the slide ranged left (fig. 7.15a), centred (fig. 7.15b) or ranged left with bullet points, lower case (fig. 7.15c), according to taste.

For optimum legibility in the auditorium, the first row of seats should not be nearer than twice the longest dimension of the projected slide and the last row of seats should not be further than six times the longest dimension of the projected slide. Thus, if the longest dimension of the projected image is 2.5m, the nearest person should be no nearer than 5m and the furthest no further than 15m away. As an incoming speaker you are unlikely to know much about the lecture hall. So, as a rule of thumb, you can reckon that letters on slides will be legible on the screen if they are big enough to be read when you hold them up to the light. Don't forget though that the amount of light reflected off the screen declines by the square of the distance to the viewer (the "inverse square" law), so try to avoid dark slides.

Tables

As mentioned above, tables are used to provide data, often raw data, for the reader to study at leisure and, if necessary, work on himself. Thus, tables taken from papers are seldom appropriate in oral presentations. If actual data are an essential part of your argument, tables from papers can be simplified to include only the data specifically applicable to the present argument. The constraints of slide making, in terms of how much you can reasonably get on a slide, apply.

Charts and graphs

Effective charts and graphs from your papers can be transferred easily to slides or OHPs to convey all the relationships between quantitative variables described above. They are particularly effective in depicting clinical progress (fig. 7.13). All the limitations described above must be taken into account to avoid redundant and misleading material and to ensure legibility. If only one aspect of a paper is being presented, it may be that a chart or graph would be better redrawn to remove "data-ink" that is not relevant to the current presentation. This is better than asking the audience to ignore parts of a slide or expecting them to concentrate on a small part of it. The problems of legibility, particularly type size and thickness of lines, and vertical presentation must particularly be considered. It is sometimes advantageous to convert graphs from the "portrait" format, appropriate for upright journal pages, into the "landscape" format appropriate for slides by elongating the x axis and reducing the y axis, if this can be done without distorting the data, although the speaker will be there to prevent misunderstandings.

Sometimes the charts and graphs prepared for conference presentations may have to be converted for publication in papers. In this case the reverse considerations apply to the figures for the paper, which can have more detail and for which the type sizes need reconsideration.

Photographs

Photographs should really be relevant to the message or the proof of the presentation. Many authors slip in pictures of famous past contributors to the subject, and sometimes show totally irrelevant artworks "to give the audience a bit of light relief" and to confirm the polymathic nature of the author's mind, together with his erudition. You may think you can get away with the most awful photographs – too small, low magnification, out of focus, distorted colours – because they are on for a short time and you talk through their time on the screen. They may be justified because they are the only ones there are or may ever be, but they leave a subliminal bad taste, whereas presenters with stunningly appropriate photographs are never forgotten.

Visual displays in posters

You study visual displays in papers wherever you like, usually sitting or even lying down, for as long as you like, and you can copy them for future reference. You have to view projected transparent visual displays in oral presentations at the same time as listening to the speaker, for as long as he leaves them on the screen. You can stand and look at the printed displays in posters for as long as you like while they are on display. In fact the whole poster is a composite visual display and must be treated as such. So you want the message and a bit of proof of a piece of work that is fairly easy to understand and fairly conclusive (Chapter 20).

The poster differs from the paper and the slides in that all the components must be legible from 1 to 2 metres (3 to 6 feet) away, with the title legible from 4 to 5 metres (12 to 15 feet) away to attract distant passers-by. It is a good idea to use standard (British Standard 4000) size sheets, A5 (148 x 210mm, 5.9 x 8.3in), A4 (210 x 297mm, 8.3 x 11.7in) or A3 (297 x 420mm, 11.75 x 16.5in). If you fold an A3 sheet in half along its long side, you get the A4 size, and folding again gives you the A5 size. Foolscap or other sizes, as used in the USA and some other countries, are difficult to scan or photocopy on equipment made for the "A" standard.

As a guide for you to try out and modify for your own ideas, Simmonds and Reynolds (1994) recommend:

- Titles: 48pt bold on 54pt linefeed (Times Roman, a serif typeface)
- Names: 24pt italic on 30pt linefeed (Helvetica, a sans serif typeface)
- Addresses: 18pt italic on 22pt linefeed (Helvetica)
- Subheadings: 24pt bold on 30pt linefeed (Helvetica)
- Text: 18pt on 22pt linefeed (Times Roman)
- Print on A4 paper and enlarge on the photocopier by 141%
- Laminate the sheets to preserve them and keep them clean
- Take "Velcro" pads to fix them to cloth boards, double-sided tape for hard boards

I recommend the use of minimum text, mainly for the message (in the title and conclusions) and a little background, with photographs and graphics as supporting data. A two-colour format, such as black and red, on a white background is attractive.

References

Asher R. A woman with the stiff man syndrome. British Medical Journal 1958;i:265–6.

Hillaby JD. Journey through Britain. London: Paladin, 1970.

James DW, Whimster WF, Hamilton EDB. Gold lung. British Medical Journal 1978;1:1523–4.

Reynolds L, Simmonds D. Presentation of data in science. The Hague:

Martinus Nijhoff, 1983.
Reynolds L, Simmonds D. Computer presentation of data in science. Dordrecht: Kluwer, 1989 (based on Apple Macintosh).
Simmonds D, Reynolds L. Data presentation and visual literacy in medicine and science. Oxford: Butterworth-Heinemann, 1994.
Snowberg RL. Bases for the selection of background colours for transparencies. Audio Visual Communication Review 1973;21:191–207 (quoted by Reynolds and Simmonds 1983, p. 102).
Tufte ER. The visual display of quantitative information. Connecticut: Graphics Press, 1983.

C
WRITING

8. Planning the writing

A naturalist's life would be a happy one if he had only to observe and never to write.
Charles Darwin (1809–1882)

How the results will be published should be at the back of the research worker's mind all the time he is planning and carrying out the research, thus reducing the new effort required when the work is finished and the writing starts.

As he does his literature searches (Chapter 2) he notes those journals which are most interested in his field and in tune with his ideas, the range of their interests, their instructions to authors and article layout, the quality of the figures, tables and illustrations, who the editor is and who is on the editorial board. Are they general or specialist journals? How general, how specialist?

References to potentially quotable articles from the literature searches can be stored, completely referenced as for a reference list, in the computer against writing-up time. It is also useful to keep a photocopy of the complete article for final checking, both the reference details and what they actually say. If you are looking up references on the databases, you can dump them on to disk or on to your computer in the form that they have been entered into the database. There are, however, computer programs, such as Reference Manager,[1] with which it is possible to format the references according to the requirements of any of the journals in the system. This may already be available in your library (Chapter 4).

Selection

Rather than too little to write about, the research will almost certainly produce too many observations, too many experiments and too much data to be published in one or even several papers. The details of what you did and the data

1 Reference Information Systems, Camino Corporate Center, 2355 Camino Vida Roble, Carslbad, CA 92009-1572, USA

obtained should have been stored in a retrievable form as you went along, otherwise details will be missing or beyond recall when you need them for the writing. Some data will have to be omitted, so which data are relevant to the first article? Selection is a very important part of clarity. You have to think in terms of what messages you have to convey and what supporting evidence you need for them.

You should also be thinking about who will be your co-authors on the papers and obtaining their agreement. The Uniform Requirements (Appendix 1) say that any author should be able to defend the paper if the other authors are suddenly incapacitated. This is surely impractical, since co-authors are often specialists in different fields who have collaborated on a project. It is not reasonable to expect a physician, a biochemist, a radiologist and a histopathologist to be able to explain the significance of each other's contributions. But co-authors should have made a definite contribution to the work. In these days of quantitation of published output for reasons other than the sharing of biomedical knowledge for the benefit of all, gift authorship to those who have made only token contributions is condemned (Appendix 1). The author who organises the paper and does the bulk of the writing is usually the first author and takes the most credit, the senior author usually goes last, and there is no specific pecking order in between, except that when the paper is cited the first three are named followed by *et al.* (abbreviation for the Latin *et alia*, meaning "and the others") (Appendix 1).

Marketing, or targeting the journal

Before you get too deeply into the writing but after you have thought out your message, it makes sense to select the most suitable journal at which to aim your message. Should it be a general journal with a readership of many audiences or a specialist journal with only a few? For example, "*Cohort study of predictive value of urinary albumin excretion for atherosclerotic vascular disease in patients with insulin dependent diabetes*" could be expected to appeal to those interested in cohort studies, renal excretion of albumin, atheroma and diabetes, whereas "Cell cycle distribution of rat thymocytes" would have a more specialist audience. You should choose a journal that you know to be interested in your field with standards of photographic reproduction and layout that suit your material. Is the biomedical work of the editor and editorial board members known in your field? Will they find your message interesting or antipathetic? You should study the articles in previous issues and the instructions to authors. What types of paper does the editor publish (see pp. 136–8)? How long are the original articles? Does he accept short reports? Does he publish only review articles? You should regard your choice of journal as a form of marketing (Albert 1996). Companies do not just make things and then hope someone will buy them; they find out what the public wants and then market those – usually with some form of publicity or advertising so that the public knows they are there. Now there's a thought for our research worker!

In other words you should remember at this stage that your aim is to be accepted for publication, and that the people you have to convince that the paper is worth publishing are the editor, and then his referees. Nobody else matters at this stage: not yourself with your altruistic ideas of the purity of science and communicating with your fellow scientists; not your boss; not the various internal referees you may consult; and probably not even your co-authors with their differing agendas. They will forgive you if you are successful.

Does your target journal accept papers in Vancouver style (Appendix 1)? This is the most authoritative set of instructions to authors, because unlike any other it is supported by a large number of biomedical journals. It is well worth keeping at your bedside to reread when all other entertainments fail. It sets out a position for virtually every eventuality, but not everyone agrees with all their solutions. You must therefore also read the instructions to authors of your target journal. My comments on the components of the structure of an article in Chapter 9 represent a personal relationship with the Uniform Requirements.

What is your target journal's "impact factor"? The impact factor is one of the brainchildren of Eugene Garfield, and is part of his "citation science" (1979). It is essentially a way of judging the importance of a journal by counting within a year the total number of times the articles published in it in a previous year have been cited in all other journals in the database (in Garfield's case, Current Contents), divided by the number of articles available for citation. This approach was only made possible by the advent of computerised databases in which each citation of each paper in each journal is recorded.

A journal's impact factor is taken, rather illogically, to be a measure of quality, and so biomedical workers struggle to be published in high-impact-factor journals. This approach takes no account of why a paper was cited, nor the types of papers – for example, a methods paper, a review, or a case report, although one might be expected to be cited more often than another; and no account of whether the journal is general or specialist. Your personal impact factor can be obtained either by simply giving each of your papers for the year the impact factor of the journal it was published in and adding the whole lot up, or, over time, the number of citations of your paper can also be extracted and manipulated.

The use of such factors as "performance indicators" has generated much heat in the research and academic world (Mullen, 1985; Chapman, 1989; Glen, 1990). The biomedical research worker should use them to his advantage if possible. This debate, essentially about the best use of public money, continues.

Sit down and do it

When the time to write comes the author must sit down and do it. Research workers often feel that they have been trained to do research and, at the writing stage, regress to the frustrations experienced with school essays. But this sort of writing is different; it is purposeful and can be exhilarating when you get into it. Many people, including myself, keep putting it off because they find getting started difficult, slow and daunting. This is because the initial stage involves thinking and organising their thoughts. You may be told to "work out what you want to say", but that is not helpful to me, as I usually do not know what I think until I see what I have said. A better instruction is to "think out what your message is", and that will help with targeting and marketing.

But this is still just putting the writing off. Where should you start? Obviously you have to have a clear idea of the overall structure of your paper, which is likely to be that set out in Chapter 9, in the context of your target journal and its Instructions to Authors, although you may have an eye to other types of writing, as set out in Chapter 12.

Word-processing

Most of the otherwise excellent advice in the books on medical writing (Hawkins, 1967, Woodford, 1968, Thorne, 1970, Fishbein, 1972, Calnan and Barabas, 1973, O'Connor and Woodford, 1975, Lock, 1977, O'Connor, 1978) comes from past times when each draft depended on a retype, a severe limitation. This was before we had the amazing word-processor revolution, i.e. before the early 1980s. Even recent books make little comment on it. One chapter (mine) of Hawkins and Sorgi (1985) was word-processed. Barnes (1986) was enthusiastic, however, commenting that, from the ancient Greeks onwards, "writing owes its existence to technological advances". Day (1989) merely mentions word-processors, while Turk and Kirkman (1989) argue that "the new technology changes the *processes* of writing, not its principles". Goodman and Edwards (1991) have little to say about their contribution. O'Connor(1991) suggests rather negatively that "word processing can seduce you into thinking you have written a masterpiece when all you have done is written a few paragraphs and moved them around a few times", and adds that it is important to make back-up copies. Hall (1994) refers to the use of computers in connection with statistics and databases, while Simmonds and Reynolds (1989, 1994) deal primarily with data presentation.

Before the advent of word-processors one already had an enviable independence from other fickle and expensive people if one could type. With a word-processor you *have* to type, and it is well worth teaching yourself to touch type. With this liberating tool you can insert ideas as they come to you and move them easily into different sequences, so that I suspect most people, like myself, are in a constant state of redrafting. "How many drafts should I reckon to go through?" has become a meaningless question.

Nevertheless I still potter around, sometimes for days, pushing my data and reference materials around my office until suddenly I'm off. There are many ways of proceeding. Some people prefer to start at the beginning and plough through to the end. Others prefer to start with the parts they are clearest and most knowledgeable about, such as what they actually did (Methods) or what they found (Results). Then they have a chunk of the work in the can, as film makers say, before tackling the more problematic parts of why they started (Introduction) and what they think it means (Discussion); and then putting it together like a jigsaw puzzle. Either way works perfectly well, with rethinking and polishing afterwards or, indeed, as you go along. For the principles of *re*writing, however, see Chapter 10.

References

Albert T. Winning the publications game. Oxford: Radcliffe Medical Press, 1996.

Barnes GA. Write for success: a guide for business and the professions. Philadelphia: ISI Press, 1986.

Calnan J, Barabas A. Writing medical papers: a practical guide. London: Heinemann, 1973.

Chapman AJ. Assessing research: citation-count shortcomings. The Psychologist: Bulletin of the British Psychological Society 1989 August:336–44.

Day RA. How to write and publish a scientific paper. 3rd edn. Cambridge: Cambridge University Press, 1989.

Fishbein M. Medical writing: the technic and the art. 4th edn. Springfield: Charles C Thomas, 1972.

Garfield E. Citation indexing. Its theory and applications in science, technology, and humanities. Philadelphia: ISI Press, 1979.

Glen JW. Bibliometric performance indicators: how their performance can affect the work of editors. European Science Editing 1990;41:3–4.

Goodman NW, Edwards MB. Medical writing: a prescription for clarity. Cambridge: Cambridge University Press, 1991.

Hall GM. How to write a paper. London: British Medical Journal, 1994.

Hawkins CF. Speaking and writing in medicine: the art of communication. Springfield: Charles C Thomas, 1967.

Hawkins C, Sorgi M. Research: How to plan, speak and write about it. Berlin: Springer-Verlag, 1985.

Lock S. Thorne's better medical writing. 2nd edn. London: Pitman Medical, 1977.

Mullen PM. Performance indicators – is anything new? Hospital and Health Services Review 1985, July: 165–7.

O'Connor M. Editing scientific books and journals. Tunbridge Wells: Pitman Medical, 1978.

O'Connor M. Writing successfully in science. London: Harper Collins Academic, 1991.

O'Connor M, Woodford FP. Writing scientific papers in English. Amsterdam: Elsevier, 1975.

Simmonds D, Reynolds L. Data presentation and visual literacy in medicine and science. Oxford: Butterworth-Heinemann, 1984.

Simmonds D, Reynolds L. Computer presentation of data in science. Dordrecht: Kluwer, 1989 (based on Apple Macintosh).

Thorne C. Better medical writing. London, Pitman Medical, 1970

Turk C, Kirkman J. Effective writing: improving scientific, technical and business communication. London: E & FN Spon, 1989.

Woodford FP. Scientific writing for graduate students. New York: The Rockefeller University Press, 1968.

9. Original articles

Reading maketh a full man; conference a ready man; ***and writing an exact man.***
Francis Bacon (1561–1626)

It is generally unwise to try to publish your research in any format other than that of a standard original "paper" or "article" (the words are used interchangeably). This is what busy editors, referees and readers are used to and know their way around. There are other, less well-defined, formats for other types of writing (Chapter 12) but non-standard formats, such as the "stream of consciousness" in which the work is reported chronologically as it happened with all its ups and downs, are distracting for the reader, who is not used to them. Subgroups of the original article include the case report (Chapter 1) and the short report (p. 110).

The standard format has a clearcut core structure consisting of the **I**ntroduction, the **M**ethods, the **R**esults **a**nd the **D**iscussion, conveniently known colloquially as the IMRAD structure. The abstract is a separate component. Seven important peripheral components also have to be written, in the manner set out in the Instructions to Authors:

- title
- running title
- author sequence
- name and address of the corresponding author
- list of keywords
- acknowledgements
- references

Whether the IMRAD components have been written effectively can be tested by a series of test questions, known as the Bradford Hill questions (1965). This valuable technique is further developed for rewriting and indeed for reading (Chapter 10)

The Introduction

For the introduction the writer must give the reader the answer to the question "Why did you start?". In other words, "What gap in knowledge did you set out to fill?" or "What controversy did you try to resolve?" For example, "A said the world was round. B said the world was flat. We set out to resolve this question with our new machine, the mondometer." What is not needed in the introduction is a complete review of every aspect of your subject since the time of Hippocrates, although this is required in the historical section of a thesis (Chapter 12). The reader wants to be clear what biomedical hypothesis is being tested or what question answered and why you set out to answer it, because the rest of the paper is devoted to supporting the message you have derived from the result. Some writers argue that a preview of the answer to the question should come at the end of the introduction: "The mondometer showed that the world was elliptical". Others, including the Uniform Requirements (Appendix 1), think that this spoils the denouement. My feeling is that the paper is a technical presentation, not a story; that the reader may know the answer from the abstract; and that the criterion is "Does it help the reader?". If so, do it. In any case the editor will have the last word.

Methods (materials and/or patients)

You must use the term your target journal uses, but most people regard "materials" as too inanimate to include patients or people. I regard "methods" as adequate on its own.

Bradford Hill's question here is "What did you do?". To orient himself, the reader certainly wants to know first on whom or what you worked, i.e. was the work done with patients, animals, *in vitro* preparations, or other forms of biomedical materials. It is useful here to give the numbers of patients or animals or specimens so that the reader can get a feel for the amount of work that has been done and form an opinion as to whether it is likely to have been enough. Some purists argue that this is more appropriate in the results section, but it is less helpful to the reader there. How were the subjects selected?

Second, the reader must be able to understand what was actually done. For example, "automobile repair shops within one hour of Boston were visited" is inexact. By bicycle in rush hour, by helicopter? The question is "Is there enough information for the reader to repeat the work himself?" Apparatus, drugs and chemicals used must be clearly stated. For standard techniques, for example biochemical procedures, references may be given, but any modifications must be included, since it would not be possible to repeat the work without them.

Third, statistical methods used should be included, with an explanation as to why they were chosen. For example, parametric tests for normally distributed data. "Avoid sole reliance on hypothesis testing", says the Uniform Requirements. If the tests are controversial, why you used them can be dealt with in the discussion. What the statistical referee will be looking for is set out in Appendix 2.

Fourth, if appropriate, you must mention that you obtained the approval of your ethics committee, and obtained informed consent from your patients, (Chapters 5 and 11). The editor will be very sensitive to this aspect.

Results

"What did you find?" Simple, perhaps, but how do you express your findings – in the text, tables or figures/illustrations? The editor does not have space for all three. Think carefully about the results that are important for (and against) your message and express them in order of importance in the text. Put supporting data, especially raw data, into tables so that the reader can manipulate them himself if he wishes. Figure 9.1, showing manipulated data only, with no

Morphometric features (means and standard deviations).

Cells	Medullary fibrosis		P value
	group II (n = 20)	group II (n = 21)	
Frequency			
total/mm²	62.4 ± 41.4	49.6 ± 17.7	n.s.
nucleated forms (%)	73.3 ± 8.3	63.0 ± 5.4	<0.001
cytoplasm-fragments (%)	17.9 ± 6.1	21.9 ± 4.5	<0.05
naked nuclei (%)	7.0 ± 4.9	10.4 ± 2.8	<0.005
emperipolesis (%)	1.8 ± 2.0	4.7 ± 2.6	<0.001
nucleated megakaryocytes/mm²	46.0 ± 32.3	31.3 ± 12.7	<0.05
Diameter (µm)			
nucleated forms	21..3 ± 6.7	27.1 ± 4.2	<0.001
cytoplasm-fragments	18.2 ± 4.9	26.1 ± 5.6	<0.001
naked nuclei	11.8 ± 2.9	15.6 ± 2.3	<0.001
Circumference (µm)			
nucleated forms	65.7 ± 19.7	78.5 ± 12.3	<0.005
cytoplasm-fragments	55.8 ± 15.5	76.3 ± 15.1	<0.001
naked nuclei	37.6 ± 8.9	47.8 ± 7.9	<0.001

Fig. 9.1. Manipulated data only; no indications of ranges or skewedness.

indication as to the statistical method used to test significance, illustrates a reader-unfriendly table. Use graphs and figures, "*pictures to show numbers*" (p. 72) to help the reader understand the distributions and relationships within the results. Remember that the least satisfactory type of figure for *comparing* sets of data is the pie chart (Chapter 6).

Discussion

"What does it mean?" Here the author has more freedom, but he must not forget that the reader needs to be able to find his way around the verbiage. Thus he is helped by an initial paragraph summarising the results that are to be discussed, preferably in the sequence already chosen as the order of importance in the text of the results, plus other aspects, preferably relevant, that the author wishes to discuss. Readers like sequences, especially if they remain constant. Each topic can then be discussed in turn in its own paragraph, or, if longer, under its own subheading. The topic can be discussed for itself and in relation to other work in the same field. Then the reader is helped by a final summary paragraph which incorporates the message and a summary of the evidence that the message may be believed, because that is where readers expect to find these two essential items.

Abstract

Unfortunately for abstract writing, many authors do not understand that, as the number of journals and papers to be read increased, editors thought that readers would like to have a summary at the beginning of the paper. Some editors wrote it themselves and called it a "synopsis", others hijacked the summary paragraph at the end of the discussion and some called it an "abstract". These were collected into "abstract journals", and then into computer databases. The space available in the data bases meant that abstracts were confined to a certain number of words, usually 150 (but see your target journal's Instructions to Authors), with any excess omitted. Clearly those who wished to see only the abstract of an article nevertheless wanted to know "Why did you start? What did you do? What did you find? What does it mean? What is the message? Can I believe it?" So the answers had to be compressed, like a telegram, into 150 words. The abstract is by now a separate entity from the main paper and has to be read as such. Thus those who read the main paper need a proper summary paragraph at the end of the discussion.

Abstract
Objectives—To estimate the risk of having twin infants for mothers who are twins; to investigate the genetic influence on twinning.
Design—Retrospective study of multiple births in two nationwide registries.
Setting—Sweden.
Subjects—Multiple births among 31 586 deliveries between 1973 and 1991 to women who were twins.
Main outcome measures—Numbers of monozygotic and dizygotic twin births expected and estimated.
Results—Women who are dizygotic twins have a moderately increased risk of having twins (relative risk 1·30, 95% confidence interval 1·14 to 1·49) which seems to be completely the result of dizygotic twinning. When a mother is a monozygotic twin, her risk of having twins of the same sex is significantly increased (1·47; 1·10 to 1·97). This is the result of an excess of monozygotic twins (39 pairs estimated, 18 expected).
Conclusions—Women who are twins have an increased risk of giving birth to twins. Genetic components of monozygotic and dizygotic twinning seem to be independent.

Fig. 9.2. The structured abstract lists information under a series of headings. (Courtesy of *BMJ*)

The structured abstract

Commercial organisations did not like the fact that, although included in the main paper, important aspects of trials of their products could easily be omitted from the abstract, with adverse commercial effects, Hence the structured abstract emerged with information under a series of headings (fig. 9.2, the exact list and word count of which for any journal using structured abstracts, is obtained from the Instructions to Authors. Usually such abstracts may allow up to 450 words, a considerable increase, which reflects the increased amount of space within the databases. The main advantage of the structured abstract for authors is that it makes them think hard about the contents under each heading, which illuminates the writing or rewriting of the paper itself.

Title

The title is possibly the only permissible form of publicity or advertising available to the scientific biomedical paper. Its use is controversial. How does it or should it differ from a headline? My own view is that it should be true. "Punk rocker's lung: pulmonary fibrosis in a drug snorting fire eater" (*BMJ* 1981;283:1661–2) may be catchy but does not convey the truth, which scrutiny of the paper revealed was that, although the pulmonary fibrosis might have been attributable to the drug snorting or more likely to the kerosene inhaled for fire eating, it could not be attributed to being a punk rocker.

The elegant and spacious age of "On the pathological changes in Hodgkin's disease, with especial reference to its relation to tuberculosis" (Reed, 1902) and even "On writing for the *Lancet*" (*Lancet*, 1937) has probably gone for ever. But authors get worse guidance on writing their title than on any other aspect. Uniform Requirements says that the title "should be concise but informative", which is not very informative. Hawkins (1967) says that "it can, without sensationalism, allure as well as inform". O'Connor and Woodford (1975) want "as much specific and intelligible information as you can in as few words as possible". Barnes (1986) reckons the reader likes it "if you narrow the topic and plunge deeply rather than cover a large terrain... You then hint at topic and approach". One major reason why medical journals are dull, apart from being part of your work to read, is "the list of long and boring titles", but a worker in the same field might miss a relevant paper on "first pass" if the title is elliptical or obscure (Lilleyman, 1994); there lies the dilemma.

Surely it should give some idea of the message? "Gold lung" *(BMJ* 1978;i:1523–4) may be striking but does not convey the right message; as one learns from reading the paper a comprehensive title would have been: "A case of reversible pseudocarcinomatous alveolar pneumocyte proliferation caused by intramuscular sodium aurothiomalate treatment for rheumatoid arthritis".

Then it can be used to guide readers to or away from the paper by targeting the potential audiences. For example, "Rickettsial endocarditis in community dental patients" might attract audiences of microbiologists, cardiologists, community physicians, dentists and possibly epidemiologists. Multiple audiences such as these might justify a general target journal (with a bigger impact factor!), rather than a specialist microbiology journal. But will any of these audiences be disappointed? What is the message? "Rickettsial endocarditis does not occur..."? "...does occur..."? "...occurs in special circumstances..."?

Why cannot we be told without having to go to extra effort? Our reading of the biomedical "literature" is work and time, not leisure reading. I believe that readers need a verb in the title, such as a newspaper headline usually has, and that to be meaningful it should convey the message, as in: "Rickettsial endocarditis is not a rare complication of congenital heart disease in community dental practice: a report of five cases". The reader then turns to the main paper for the evidence to support the truth of the message. The Instructions to Authors may indicate a certain maximum number of characters, including spaces, for the title.

Set out the cover page with the title, authors, author affiliations, running title (or running head), and author and address for correspondence, as instructed in the Instructions to Authors, noting also the appearance of these components in the original articles in the journal, since editors may request material in one format but may prefer, usually for historic reasons, to print it in another, making the necessary changes in house.

The running title

A running title is the short title used for page headings. It should be supplied if asked for in the Instructions to Authors, and should not exceed the number of characters (including spaces) if specified. An example of this is: main title "Resistance to recombinant human erythropoietin due to aluminium overload and its reversal by low dose desferrioxamine therapy", with running title of "Resistance to recombinant human erythropoietin" (46 characters).

Authors

Authorship has already been discussed. Having decided on your co-authors, you must check the spelling of their names, their initials or first names, their qualifications if required and their affiliations, as specified in the Instructions to Authors, noting also the appearance of these components in the original articles in the journal. For example:

> M Yaqoob, R Ahmad, P McClelland, KA Shivakumar, DF Sallomi, IH Fahal, NB Roberts,[1] T Helliwell[2]
>
> Departments of Renal Medicine, [1]Chemical Pathology and [2]Pathology, Royal Liverpool University Hospital, UK
>
> Correspondence: M Yaqoob, 6Z Link Unit, Royal Liverpool University Hospital, Prescot Street, Liverpool L7 8XP, UK

Note how the affiliations are related to the authors. Note also that you must ensure that all authors are happy to be authors, and, if required by the Instructions to Authors, sign the covering letter to the editor. Furthermore they must sign again for any subsequent modifications to the paper, for example after refereeing. Authors may also be required to state the source of any funding they have had and some funding agencies insist that funding must be acknowledged in the paper. Authors may also be required to state that there is no conflict of interest (see Appendices 1 and 2); for example, that you do not have vested interest in a drug under investigation in your paper.

Keywords

The Instructions to Authors will show whether these are required as part of the retrievability system for biomedical papers. To obtain the 3–10 short words or phrases, you should choose them from the list of Medical Subject Headings (MeSH) of the *Index Medicus*, not including words that are already in the title. Retrievability for your paper is important, not least because it contributes to the number of times it will be cited, and is discussed in Chapters 4 and 8.

Acknowledgements

Between the end of the discussion and the references it is usual to acknowledge contributions to the work, including the paper, from anyone who is not an author. One must be reasonably selective about this and also, as with co-authors, ensure that those acknowledged are happy to be associated with your paper (Appendices 1 and 2).

References

You provide references to indicate the sources from which you have obtained your information, as well as for support and authority. But the reader wants to know your thoughts about the information you quote, especially in theses, not just what the original authors thought. Don't think that the value of an article is measured by the number of references; they should not be included merely to show erudition. I value the reference lists, and, when refereeing, often wish to photocopy for future use. I am never sure whether this is allowed within the rules of the International Committee of Medical Journal Editors (Appendix 1); presumably the list belongs to the author.

Garfield (1982) clearly expressed the value of references: "The importance of citing sources in scientific publications should not be taken lightly. After all, citations are the reward system of scientific publication. To cite someone is to acknowledge that person's impact on subsequent work. Citations are the currency by which we repay the intellectual debt we owe our predecessors. Furthermore, failing to cite sources deprives other researchers of the information contained in those sources and may lead to duplication of effort."

While you are writing the references into the paper it is easiest, and indeed most reader-friendly, to use the Harvard system in which the names the authors (up to two; otherwise the first and *et al.*, and the dates of the publication in the text; then make the reference list alphabetically by first author (Example 1). The Harvard system uses a lot of space and the Instructions to Authors may insist on you numbering the references in the text and listing them by consecutive number (Example 2). This has the disadvantage that references are hard to find in the list, so you may have to number the alphabetic list sequentially and then insert the numbers back into the text, where they will no longer be sequential. The examples show the punctuation approved by the International Committee of Medical Journal Editors (Appendix 1). Indeed, Uniform Requirements gives the most comprehensive list of ways to put a reference, very well worth having by you if you wish, for example, to reference a computer file or a map. Once again check the format and punctuation required by the target journal's Instructions to Authors, and note how they appear in the text. As before, the journal may request or accept references in one format but print them in another. Also check all references against your original photocopies before sending the paper off. I have sometimes found 15–20% of mine to contain errors, of which perhaps 3% were completely irretrievable. It is also worth checking to ensure that the references you quote do say what you say they say; some of them may say the opposite (de Lacey *et al.*, 1985).

Example 1

Jones and Bloggs (1962) reported hypercalcaemia in three cases of Upps syndrome, although hypocalcaemia has also been reported (Asquith and Tate, 1965; Meredith *et al.*, 1971)

Reference list

Asquith AB, Tate TT. Hypocalcaemia in Upps syndrome. In: Smith J, ed. Upps syndrome abandoned. London: Castle Moat, 1965:96–143.

Jones J, Bloggs B. Hypercalcemia in three cases of Upps syndrome. J Calcemia 1962; 84:223–7.

Meredith MM, Valve PP, Wells NV. Hypocalcaemic neuropathy: a complication of Upps syndrome. Acta Calcaemia 1971;2:34–8.

Example 2

Jones and Bloggs[1] reported hypercalcaemia in three cases of Upps syndrome, although hypocalcaemia has also been reported. [2,3]

Reference list

1. Jones J, Bloggs B. Hypercalcemia in three cases of Upps syndrome. J Calcemia 1962; 84:223–7.
2. Asquith AB, Tate TT. Hypocalcaemia in Upps syndrome. In: Smith J, ed. Upps syndrome abandoned. London: Castle Moat, 1965:96–143.

3. Meredith MM, Valve PP, Wells NV. Hypocalcaemic neuropathy: a complication of Upps syndrome. Acta Calcaemia 1971;2 :34–8.

Example 3

Jones and Bloggs[2] reported hypercalcaemia in three cases of Upps syndrome, although hypocalcaemia has also been reported.[1,3]

Reference list

1. Asquith AB, Tate TT. Hypocalcaemia in Upps syndrome. In: Smith J, ed. Upps syndrome abandoned. London: Castle Moat, 1965:96–143.
2. Jones J, Bloggs B. Hypercalcemia in three cases of Upps syndrome. J Calcemia 1962; 84:223–7.
3. Meredith MM, Valve PP, Wells NV. Hypocalcaemic neuropathy: a complication of Upps syndrome. Acta Calcaemia 1971;2:34–8.

References to conference abstracts can be listed. References accepted for publication can be listed as "in press"; those "in preparation" or "personal communication" can only be mentioned in the text, as they are not retrievable. If a reference is totally unobtainable it can be cited as "Smith (1832), cited by Adams (1973)", with Adams's reference. Then the responsibility for what Adams said Smith said is down to Adams and not to you. No one can remember the formats for every type of reference from journals, books and every other retrievable item, and your target journal may not be very comprehensive, but the Uniform Requirements has all 35 in Appendix 1. Uniform Requirements also requires the titles of journals to be abbreviated in the style used by *Index Medicus* and expects inclusive page numbers to be given in abbreviated form, as shown in the example above, 1971;2:34–8 - rather than 34–38. Neither of these requirements is author- or reader-friendly but, overall, they save quite a lot of journal space. I write mine out in full and do the abbreviating later.

Writing

You are now in a position to start writing within the IMRAD structure: you know the questions to be answered in each section; you know what the message is. Do not worry about style or grammar. Start with the section you feel most comfortable with and tap it roughly into your word-processor. You can fill in the details of the references, tables, figures and so on later. On this basis I can put in 400–500 words an hour, including pauses for thought, so a normal length paper of about 2,000 words can be done in 4–5 hours. To put this quantity of words in perspective, as a student you must have written numerous 30-minute examination essays of two to three A4 sheets (about 600 words).

Before word-processors arrived, your 2,000 words would have had to be typed as a first draft and further drafts would follow. Now, however, you can fiddle with the text endlessly, but successive printouts for review by yourself or others are really successive drafts, and as such come under the heading of rewriting (Chapter 10).

Short reports

Some journals take original articles in the form of short reports or brief communications with specific criteria. The example in Appendix 2 specifies a strict limit of 600 words, one table or illustration and five references. Although roughly following the IMRAD structure, they are not particularly reader-friendly, because there is no fixed position for the message, no abstract and usually no summary (message) paragraph at the end.

Style

Many biomedical authors and readers, coming, as they have done, from an educated background, worry that the journal format tends to take any individuality out of the articles published. Some worry that this makes their paper uninteresting, although whether or not they interest any particular reader depends upon the reader's interest and need, not those of the author. The reading biomedical workers have to do is both voluminous and in the nature of work, even if it is largely undertaken out of working hours. Some authors do manage to express themselves in a lively turn of phrase, and some manage to be much more direct and easy to apprehend than others. But we are often dealing with very intricate and complicated ramifications of biology and medicine; whether we are interested enough to read it is more a function of ourselves than of the text. I suggest that if it is boring it may be a consequence of the reader's lack of motivation as much as one of the author's writing. But the author can help by thinking always of the reader (especially the editor and referees, who will read it much more diligently than anyone else ever will), sticking to short sentences, and putting the writing into the active voice. Even in 1937 "the first person singular – the naked I – [was] no longer thought immodest" (Lancet, 1937). There are plenty of style books, most of which I find boring as they are so opinionated and unscientific, but Sir Ernest Gowers's The complete plain words, written at the invitation of the Treasury, is well worth dipping into. Finally, you may be helped by Appendix 7, which gives examples of stylish (and unstylish) use of words.

References

Barnes A. Write for success: a guide for business and the professions. Philadelphia: ISI Press, 1986.
de Lacey G, Record C, Wade J. How accurate are quotations and references in medical journals? British Medical Journal 1985;291:884–6, and correspondence pp. 1282, 1420, 1421, 1580.
Hill AB. The reasons for writing. British Medical Journal 1965;ii:870.
Garfield E. The ethics of scientific publication: authorship attribution and citation amnesia. Current Contents 1982;25:6–10.
Gowers E. The complete plain words. Harmondsworth: Penguin Books, 1962.
Hawkins CF. Speaking and writing in medicine. Springfield: Charles C Thomas, 1967.
On writing for the Lancet. Lancet Supplement 1937; January 2: i–iv.
Lilleyman J. Titles, abstracts and authors. In: Hall GM ed. How to write a paper. London: British Medical Journal, 1994.
O'Connor M, Woodford FP. Writing scientific papers in English. Amsterdam: Elsevier, 1975.
Reed D. On the pathological changes in Hodgkin's disease, with especial reference to its relations to tuberculosis. Johns Hopkins Hospital Reports 1902;10:133–96.

10. Rewriting

So that the jest is clearly to be seen,
Not in the words – but in the gap between.
William Cowper (1731–1800)

The number of successive drafts/word-processing printouts varies greatly from author to author and from paper to paper, and is of no significance provided the final product is successful. But the author will be greatly helped if he has a clear idea of what he is trying to do when he is rewriting, so that he can do it purposefully and quickly.

He is trying to remove distractions. A distraction is anything in the writing that diverts the reader's mind from the straight path planned for it by the author. Distractions leave unanswered questions floating in the reader's mind and waste his time (see fig. 10.1).

There are four types of distraction:

- structural
- grammatical
- flow/logic
- numerical/statistical

Structural distractions

Structural distractions are those caused by not sticking to the IMRAD (introduction, methods, results and discussion) structure. They are detected by what I think of as a pathologist's "naked eye" or "macroscopical" examination. The kit of tools for you to test whether your writing is in order consists a series of questions (Appendix 8). Is there a clear answer to the question "Why did you start?" in the introduction? To "What did you do?" in the methods? To "What did you find?" in the results? To "What does it mean?" in the discussion; not forgetting the refinements around these questions described in Chapter 9. So often the reader has to waste time hunting for the answers because they are in

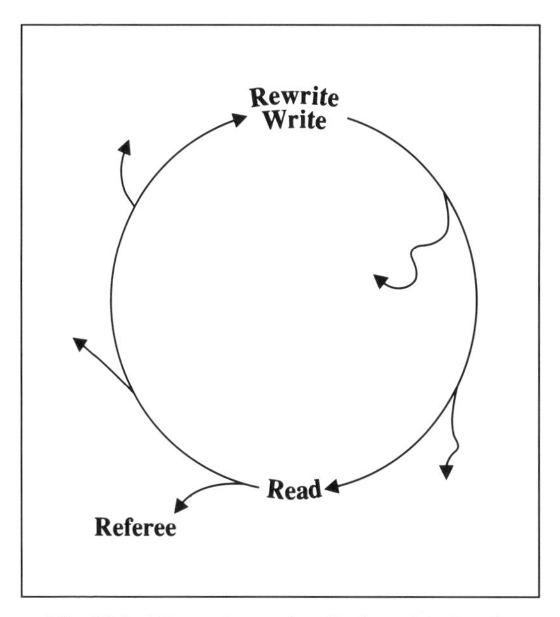

Fig. 10.1. Distractions to be eliminated during the write–read–rewrite cycle.

the wrong section or not answered at all. Then can you answer the two fundamental questions "What is the message?" and "What is the proof that the message is true?". Finally, are the answers to any of the questions in the wrong sections? Is there any redundant writing that does not contribute to answering the test kit questions? If so, cut it out.

Journal clubs are splendid fora for members to practise applying the test kit. When selecting papers for presentation to the club, firmly reject any in which you cannot answer the test questions; use the answers to the questions as the basis for your presentation, which can thus be brief and to the point. Of course it is not easy to apply the test kit to papers in other formats, such as review articles, but they are very difficult to distil for journal club presentation anyway.

Then the abstract, whether structured or not, should be checked to ensure that it contains the answers to all six questions, and indeed, if structured concisely, fills the headings required, within the number of words allowed (see the Instructions to Authors of the journal). Abstracts should not include information that is not in the text of the paper.

You should take another look at your title to ensure that it defines the topic of the paper; why it is of interest; whether it is clinical or experimental work; and the audiences who may be interested, for example cardiologists, microbiologists, pharmacologists, epidemiologists, dentists. Possibly, if you share my views, you will like to have the message in the title, not forgetting that that may be too radical for the target journal.

Have you written the rest of the peripheral structure according to the instructions?

- authors names
- affiliations
- running title
- name and address of the corresponding author
- keywords
- acknowledgements
- references

Grammatical distractions

Most people have forgotten most of what they knew about grammar and are not keen to study it now, but many readers are distracted if the grammar is wrong, often without realising why, and may misunderstand the meaning.

Detecting grammatical distractions calls for "microscopical" examination, of every word in every sentence. There are only eight parts of speech (in three groups) for a word to be: the noun, article, pronoun, adjective group; the verb, adverb group; and the preposition and conjunction group for sticking the parts together. More about these can be found in Gowers (1962, 1968).

Any author should be able to test each sentence with the following kit of nine tests:

- What is the subject and verb of each sentence and are they both singular or plural, i.e. do they "agree"?
- Is the most important noun the subject? For example, to say "The experiment was conducted by the scientist" makes the experiment more important than the scientist, whereas "The scientist conducted the experiment" makes the scientist more important than the experiment. Which do you mean?
- Is the verb in the right tense, usually the past, since all that is being reported, including the writing, has been done by the time it is published. The present tense implies a universal truth. For example, "The blood pressure rises during stress". Is this true in all circumstances, everywhere? Is this what you mean? The future tense is seldom the right tense.
- If the verb is passive, would the sentence be clearer in the active, and is information being withheld? For example, if "The experiment was carried out at –20°C", who carried it out?

- Is it clear what any pronouns stand for? For example, "They reacted unfavourably"; who or what did?
- What noun do any adjectives or adjectival clauses or phrases tell you more about? Could they be left out? This test includes simplifying noun clusters such as "daily male patient liver enzyme status", where nouns are also used as adjectives. This test should also help the author to rewrite unattached or dangling clauses or phrases, such as "He was hit by an umbrella belonging to an old lady with whalebone ribs".
- What verb do any adverbs or adverbial clauses or phrases tell you more about? Could they be left out?
- Is each preposition used sensibly in relation to time (since, while), place (where) and person (who, whose)? If time is not implied, "Since all tumours are visible at bronchoscopy..." should be "As all tumours are visible at bronchoscopy...". What about "Where necessary...", "When necessary..." or "If necessary..."? Is place or time or neither implied?
- Finally, one aspect of punctuation is important. Do any commas help or hinder the sense? For example, in "The patients, who were on the waiting list, were soon seen...", the phrase "who were on the waiting list" is merely a comment, whereas in "The patients who were on the waiting list were soon seen...", the same phrase defines the group of patients. If you use only one comma, "The patients, who were on the waiting list were soon seen...", the reader cannot tell if you intend to comment or define, and must ask you or be left with a distraction.

Authors who find a grammatical distraction should not worry too much about exactly what the trouble is, but should just rewrite the sentence in another way to get round the distraction as simply as possible

I think the attempts in most medical writing books to help biomedical authors weed out grammatical distractions are unsuccessful. The readers of these books are inundated with points and examples that they cannot convert to their own use without embarking on a new student career in the English faculty. Furthermore, authors know that they can usually trust their target journal staff to sort out the grammar or to ask if they cannot understand it. Papers with calamitous grammar are seldom rejected on that ground alone. So why should the authors bother? Perhaps it's fortunate that some of them do, especially those whose first language is not English (Chapter 13). For those who do, enjoyable but serious and constructive help is at hand in Goodman and Edwards's (1991) 190-page book. Among many good ideas, the authors have been bold enough to set out 35 short passages for the reader to rewrite, together with their own ideas for rewriting them. If you have time for a scholarly account of why English grammar is so full of anomalies, try Palmer's Grammar (1971). If you want to understand how to turn English to your advantage, read Thouless's Straight and Crooked Thinking (1953, first published

1930). At least with these approaches you will get some fun out of a too often condescending and po-faced subject.

Flow and Logic

Authors must think out a fluent and logical way to present their information and interpretations if their messages are to go straight into the reader's mind. If the reader knows that the IMRAD structure has been used, his mind can be regarded as already formatted, like a computer disk, for flow and logic.

The test kit for testing flow and logic has five components:

- attend to long sentences and long words
- abbreviations
- elegant variations
- sequencing
- paragraphing

Long sentences and long words

Flow is obstructed by long sentences and long words. Look again at each sentence with more than ten words. Are all the words necessary? Could it be made into two sentences? Look again at each word of more than two syllables; could it be replaced with a one or two syllable word? Pomposity comes about by using long words and long sentences to make the writing (and the author) appear important (Appendix 7). The Campaign for Plain English has long tried to simplify and render public, governmental and bureaucratic pomp readable and understandable by the masses. Writing pompously can be a useful technique but should be used consciously (see example below) and for a purpose (such as concealing negative results when seeking extension of a grant), not as a casual, reader-unfriendly, time-wasting, ego-driven distraction, or from trying to write in what is perceived as journal, academic or scientific style. The latter, understandably, particularly affects authors whose first language is not English, because they try to emulate what they read in the English language journals.

Example: Pompous on purpose?

> It is suffice to say that although substantial data have been presented demonstrating the antigenicity as well as the presence of tissue and species-specific antigens of prostatic tissue of reproduction of the various species studied the demonstration of the presence of tumours specific antibodies or for that matter circulating antibodies to prostatic

> tissue or secretions by the methods of precipitation and of passive haemagglutination in the sera of patients with benign or malignant diseases of the prostate and/or following cryosurgical prostatectomy has been despite histologic and roentgenologic observations of the remission of distant metastases in cases of metastatic carcinoma of the prostate (stage 3) following the cryosurgical treatment of the primary prostatic tumour for the most part discouraging.

This sentence can be distilled to:

> Prostatic tissue antigens have been demonstrated but circulating antibodies to benign or malignant prostatic tissue have not.

Abbreviations

Flow is also obstructed by abbreviations, which are not reader-friendly because the forgetful reader may have to keep turning back to find the meaning. Abbreviations should not be used in titles, and should be avoided in abstracts. Look at each abbreviation to ensure that it cannot be spelt out each time it appears, especially if it appears fewer than ten times. If not, check that its meaning really has been spelt out the first time it appears (and in the abstract, as that may be all that some readers see). Don't forget that different readers use the same abbreviations with different meanings: for example, SI may mean:

- soluble insulin
- serum iron
- saturation index
- sexual intercourse
- stimulation index
- Système International d'Unites

Elegant variations

Elegant variations are synonyms or apparent synonyms used to make the writing more interesting, as in "the blood pressure fell, the temperature declined, the pulse rate dropped and the respirations diminished". Unfortunately, varying elegantly raises the question of whether the author means the same thing with each synonym or is trying to say something different. It is better to stick to the same word if that is what is meant, and risk being uninteresting.

Discussion
In this study feeding was started two or three hours after surgery and continued until normal diet was possible. Full feeding was achieved quickly and was well tolerated with no excessive distension. No patients progressed to postoperative total parenteral nutrition. The functional integrity of the bowel mucosa was assessed by a differential sugar absorption test. Mannitol is absorbed transcellularly and lactulose paracellularly (40 to 100-fold lower absorption than mannitol). Both sugars are excreted unchanged in the urine. Inreased intestinal permeability allows greater amounts of lactulose to be absorbed thus raising the urinary lactose:mannitol ratio. This occurs in critically ill patients,[14] patients with multiple trauma,[15] and patients undergoing major vascular procedures.[16] Our finding that immediate enteral feeding after bowel resection seems to prevent the rise in intestinal permeability suggests a protective role, whether by providing a physical barrier or via direct metabolic and nutritive effects on the intestinal mucosa remains unknown. The observed higher protein and energy intake in the enterally fed patients may have contributed to the improved gut mucosal integrity.

Fig. 10.2. Paragraph with highlighting showing buried topic sentence. (Courtesy of the *BMJ*)

Sequencing

Readers are greatly helped by sequencing. For example, if the paper is about three tumours, an adenocarcinoma, a squamous carcinoma and an undifferentiated carcinoma, it is much easier to comprehend if the tumours are considered in the same sequence throughout the paper. Thus, if the author decides on an order of importance for his results, for example the undifferentiated carcinoma was very responsive to the treatment, the squamous carcinoma less so, and the adenocarcinoma unresponsive, he should also discuss them in the same sequence in the discussion and adopt that sequence throughout the paper. The same applies to ideas. For example, if the paper is about endocarditis, a decision should be made about the sequence in which the cardiological aspects, bacteriological aspects and antibiotic aspects are discussed and that sequence retained throughout, not only within the IMRAD sections but also within paragraphs.

Paragraphing

Within each IMRAD section the flow of the logic depends on the paragraphing. Each paragraph should therefore be inspected to ensure that it starts with

Primary dysmenorrhoea affects up to half of post-pubescent women and is one of the commonest gynaecological complaints. Combined oral contraceptives and non-steroidal anti-inflammatory drugs are often used to treat dysmenorrhoea but are not effective in 10–20% of patients[1] and are contraindicated or unsuitable for many others. There is a need for a simple and safe new treatment for this condition.

The pain of dysmenorrhoea is associated with increased intra-uterine resting pressures and peak pressures,[2] and effective treatment is associated with uterine relaxation.[3] Glyceryl trinitrate seems to relax uterine contractions in preterm labour,[4] and we investigated whether it could be used to treat dysmenorrhoea.

Fig. 10.3. Paragraph with highlighting showing topic paragraph first. (Courtesy of the *BMJ*)

a topic sentence and ends with a link sentence that introduces the next paragraph. A good test for paragraph structure is to go through a paper and highlight the sentences containing the most important information. On review the highlighted sentences in a clear paper will turn out to be the first sentences of the paragraphs. If they are buried in the body of the paragraphs the text will be unfriendly to read because the important information is preceded by qualifying or subordinate material. It is easy to reconstruct paragraphs so that the important information is in the first sentence, often by making the buried sentence the first sentence of a new paragraph (figs 10.2 and 10.3).

Especially in the discussion, subheadings for single paragraphs or groups of related paragraphs can greatly help the reader. The first paragraph of the discussion should recapitulate the findings in the sequence in which they are set out in the results, and then set out any other issues that the author wishes to discuss. The next paragraphs should discuss each issue in the same sequence. The final paragraph of the discussion should set out the author's message clearly and summarise the proof that it is true, as well as indicating where the research should go next.

Numbers and statistics

The message is usually in the writing; the proof that the message is true often lies in the numerical results and their statistical handling. The presentation of numerical results and statistics is therefore not only important but the source of five types of distraction to test for:

- multiple experiments
- what is in the text, tables and figures
- how much raw data is put in
- showing comparisons
- showing that the comparisons are statistically significant and medically or biologically important

Multiple experiments

If the work consisted of experiments on two tumours, both *in vivo* and *in vitro*, with six cytotoxic drugs, i.e. at least 24 experiments, the reader must be led very carefully through the results and discussion to the conclusions, if he is not to become completely confused. Although editors are not keen to allow "salami" slicing of results to give authors multiple credit for the same work, dividing such material into more than one paper may be the only way to convey the messages and proofs. This may call for negotiation with the editor, initially through the covering letter.

Text, tables and figures

Editors do not like to publish the results both in the text and in tables and figures. Check that the text highlights the important results in the order of importance and the order for discussion; that the tables are used to show the numerical results (so that the reader can see their distributions and make his own calculations), the manipulated results and the statistics; that the figures are used to show relationships in the results in a visual manner to help the reader.

Raw data

Check that as much raw data as are necessary for the reader to do his own calculations are given, usually in tables. Do not give only means, i.e. the results after manipulation, so that it may be impossible for the reader to know if the raw results were normally distributed, for example, and whether the statistical tests used were appropriate. This may also call for negotiation with the editor.

Comparisons

In many papers the results of different groups are compared. It is important that these comparisons should have adequate legends and be reader-friendly. The author should check that this is so, but there are no generally applicable guidelines. The best test is for the author to persuade a colleague to look through them and to ask the author about every point he finds difficult to follow.

Significance and importance of comparisons

It is also important that the statistical differences are shown clearly, with adequate explanation of the statistical tests used. For medical and biological work, confidence intervals may be much more meaningful than significance testing. Nevertheless, if P values are given, the test used, the number of cases/items included and the standard deviation or standard error must all be specified (Chapter 6).

The many other aspects of the numbers and statistics are the province of the scientific and statistical referees (Chapter 15 and Appendix 2) and outside the scope of simple rewriting.

Informal refereeing (colleague treatment)

Although a few medical and scientific authors write original articles in memorable prose, most authors, even the most senior and experienced, do not. So a colleague who understands the principles of rewriting and is prepared to act as an independent reader is an asset, especially if he will go through it with you verbally, so he can say "What are you trying to say?", and you can reply conversationally – and write it down quickly. This "colleague treatment" (fig. 10.4) is not particularly concerned with the biomedical content of the paper - constructive criticism of that should come from someone else, such as one's supervisor or departmental head. Incidentally, it is difficult for those at different levels in the hierarchy to subject themselves to the "colleague treatment" unless they are used to it or have a long-time colleague. Some other friend, perhaps not even medical, is often useful to question illogicalities and convoluted language. It is a great help to have a colleague familiar with the techniques above to go through the paper (and you can do the same for him). You should not just ask your private referees "to comment", which just allows them to show that they are cleverer than you are. Give them specific tasks: "Can you see any silly mistakes?", "Is the science OK?", "Is the English clear?", "Do you understand the message?".

Fig. 10.4. Colleague treatment.

Conclusion

One has to be very methodical to subedit one's own paper. The test kits for removing distractions (Appendix 8) provide a basic approach which each author will add to from his own experience as it grows. The checklists in Appendix 2 may help the author to know what those on the receiving end of his paper are asked to look for.

Don't omit to reread the Uniform Requirements for manuscripts (Appendix 1) and the Instructions to Authors of your target journal, both before and after you do your subediting.

Finally, check your references in the text and in the reference list. Such a check can rescue the references that are irretrievable or contain mistakes.

References

Goodman NW, Edwards MB. Medical writing: a prescription for clarity. Cambridge: Cambridge University Press, 1991.

Gowers E. The complete plain words. Harmondsworth: Penguin Books, 1962.

Gowers E. Fowler's Modern English Usage. 2nd edn. London: Oxford University Press, 1968.

Palmer F. Grammar. Harmondsworth: Penguin Books, 1971.

Thouless RH. Straight and crooked thinking. Hodder and Stoughton, 1930. Revised and enlarged edition, London: Pan Books, 1953.

11. Submitting the work

The object of research is publication.
Professor John Ziman (quoted by Stephen Lock in the "Preface" to Hawkins and Sorgi, 1985)

Publication will not be achieved unless you finally pluck up courage to send off your paper. At last, every component has been assembled and every point in the Instructions to Authors attended to. All that remains is to write the covering letter (Appendix 1) and send the package to the target editorial office. The whole paper has been printed out, (double-spaced, with generous margins to allow editorial staff room to work legibly on it) in the number of copies requested, plus at least one file copy of the whole package for yourself for when the proofs have to be checked. The tables and illustrations will be on separate sheets at the end, with figure legends on a separate sheet because they are printed separately. Photographs, marked on the back for identification (Chapter 7) are sent in the required number of sets in separate envelopes without any sort of clips.

The Instructions to Authors may allow for papers also to be submitted on computer disk (see Appendix 2 for example, and also Uniform Requirements in Appendix 1). Note the request not to use automatic page numbering, referencing or footnotes. The hard copy pages should be numbered by hand.

Appendices 1 and 2 emphasise that, apart from the paper itself, there are quite a number of statements, photocopies and signatures that may have to be enclosed, plus a covering letter.

To write an effective covering letter, one that is to open a dialogue with the editor that may go on for some time, and perhaps continue with further papers, you should give some further thought to who will receive it – no longer just **an** editor, but a **specific** editor.

A biomedical journal is generally edited by one or more biomedical professionals, well used to the working environment of their authors, often for a biomedical society or association, which may choose to spend subscription money on a loss-making journal, but will obviously prefer to be commercially successful, often with the help of advertisers. Papers are editors' life-blood. No papers, no journal: editorial, professional and possibly commercial humiliation. Some journals are more clearly commercial enterprises with less biomedical input or control; these may only take articles they have commissioned. It is well worth the author finding out the nature, affiliations and motivation of his target journal, usually from the inside front cover where the editors and others who work for the journal are listed. The editorial board listing also gives the author valuable information about who may be asked to judge his paper.

The author's credibility is established by having the letter cleanly typed on headed paper and signed by all the authors. The more specific part of the letter should consist of introducing the paper by title and asking (not demanding) that the editor consider it for publication in his journal. This should be followed by why you are submitting it to his journal and why he should be interested in it. You may wish to suggest that certain people would not be appropriate as referees giving your reasons.

Check to see if the target journal makes any charge, either a handling charge on all manuscripts submitted, or a page charge on accepted papers according to the number of published pages, or a charge for colour printing. Journals with more advertising material are less likely to make charges.

Send it all off. If you have not had an acknowledgement of receipt within a reasonable time, you may contact the office staff to make sure it has arrived. Then you must wait patiently for the refereeing process to work its way through, without badgering the office, for at least two months (Appendix 2). For monthly or quarterly journals the wait may be longer. The office will usually tell you where the process has got to if you approach them politely.

12. Other writing, including theses

The concept that a thesis must be a bulky 200-page tome is wrong, dead wrong. Most 200-page theses I have seen are composed of maybe 50 pages of good science. The other 150 pages comprise turgid descriptions of picayune details.
Robert A Day (1989)

The original article is only one of many writing formats that biomedical workers may use to publicise their research and include in their CVs. Within the journals may be found leading articles (editorials), topic reviews, conference reports, educational articles, letters to the editor, book reviews, news items, obituaries and association business, the last two requiring some ingenuity perhaps to make them work for you.

These do not have the advantage of an IMRAD structure to write to, although the Uniform Requirements (Appendix 1) has instructions about them, and there may be instructions in the target journal's Instructions to Authors, usually concerning length.

You may be moved to submit unsolicited items for publication; you may be commissioned (and even paid) to write others. The editor may seek advice for any of these from referees. In any case, check the Instructions to Authors and concentrate your writing on your target reader, the editor. Outside the journals, the principle writing formats are articles for the scientific or lay press (a tricky but potentially rewarding field), reports to the grant-awarding bodies or sponsors, and theses for postgraduate university degrees, including MSc, PhD, MS (for surgeons) and MD (for physicians) and their equivalents in the US. In Europe, some countries and some universities allow the combination of a number of published papers to form a thesis for a PhD. If the university accepts the thesis, a copy is lodged in the candidate's college or university library, whence it can be retrieved through the databases. The thesis counts as a publication for curriculum vitae and academic credit purposes.

The thesis

Your target readers for a postgraduate thesis are specifically your supervisors and examiners, with whom you will be more or less acquainted. On the other hand university regulations for postgraduate theses are much less specific than journal Instructions to Authors, although they may nowadays specify a maximum length, in terms of numbers of words, to the relief of the examiners (Phillips and Pugh, 1994). The regulations concentrate more on the academic aspects, for example that the thesis "must form a distinct contribution to the knowledge of the subject and afford evidence of originality, shown either by the discovery of new facts and/or by the exercise of independent critical power", and "show in what respects the candidate's investigations appear to him to advance the study of the subject". Nevertheless, the regulations must be obtained and studied conscientiously, for failure to focus on the goals of the regulations has led to many a thesis having to be rewritten. It is also helpful to study any related theses already lodged in the library, to see what examiners have been prepared to accept previously.

There is no prescribed structure for a thesis, although there is a British Standard (BS4821:1990) with recommendations for the presentation of theses and dissertations. The candidate can please himself, but is unwise to do so without thinking of the examiners. A normal PhD thesis in the biomedical field usually runs to about 200 A4 pages of double-spaced typescript, consisting of seven to ten chapters including tables, figures and illustrations, plus the list of references. Brilliantly original theses can be short. Nevertheless, in either case the examiners/readers want a clear pathway for their minds to follow, preferably in the form of a hypothesis that the candidate set out to test, what he did to test it, and whether the hypothesis turned out to be right, partly right, or wrong. Once again they need to be introduced to the topic, by a sensible thesis title, an abstract (see pp. 103–4), and an introductory chapter, which can be more expansive than for a journal paper but still confined to answering the question "Why did I set out to test this hypothesis?". Everything known about the topic from its inception or the time of Hippocrates is not excluded from a thesis, and may be a very scholarly, useful and interesting part of it, but it goes into a separate chapter after the introduction, called the historical background and literature review, which "should be used to substantiate and carry forward arguments and counter arguments. It should not be read as a catalogue of vaguely relevant material" (Cryer, 1996).

There may be several sets of methods (What did you do?) and results (What did you find?). They may already have been published in a series of journal papers (which may be bound for reference within the end papers of the thesis). It may be the problem of the multiple experiments writ large (Chapter 10). The candidate has to decide whether to try to present these in comprehensive methods and results chapters or to present each set within its own chapter, with introduction, methods, results and discussion in each, perhaps corresponding to the relevant published paper. If not already published, this format can form the basis for subsequent papers for publication.

The last chapter is a comprehensive discussion (What does it mean?) of the hypothesis and the results of testing it, in its own right and also, of course in relation to the work of others, culminating in a summary paragraph outlining again the message and whether it is to be believed. As the discussion is usually long, an introductory paragraph to set out the sequence of topics to be discussed and headings for each topic paragraph are almost mandatory. It is important throughout this chapter, and indeed in all the chapters, to express to the readers/examiners what *you* think about the methods, the results and what they mean, as well as what other workers have found and thought. This demands an approach that is more personal, emphatic and possibly more controversial than is usually thought to be correct for dispassionate scientific appraisal.

Some thesis regulations will also require a further page stating how much of the work has been done by you and how much by others, and also stating formally "in what respects the candidate's investigations appear to him to advance the study of the subject".

Having submitted the thesis to the examiners, in the UK it is usual to have to appear for an oral *(viva voce)* examination before an external examiner and an examiner internal to your university. In some countries there is a very formal examination by the examiners in public; in others there is no oral examination. An oral examination is often described as a thesis defence, but defending is not the same as being defensive. You should regard it as more of an academic joust with worthy opponents. To prepare for it, reread your thesis. Work out what you think are its strengths and weaknesses and think up some questions to ask the examiners. Usually they are trying to pass you with as little further labour as possible, so follow their leads and indulge their ideas.

As with journal papers the examiners (editor/referees) may pass (accept), fail (reject) or recommend pass after modification. The latter is probably the commonest outcome. On a personal note, therefore, I find it a pity that the universities often require bound copies of theses for distribution to examiners. If it has to be modified, it must presumably be un-bound, modified to the requirements of the examiners and re-bound at extra expense. The requirements may be very detailed and specific. How are they to be conveyed efficiently to the candidate if each examiner does not have a loose leaf copy to write notes on for the *viva* and/or the candidate?

References

Cryer P. The research student's guide to success. Buckingham: Open University Press, 1996.

Phillips EM, Pugh DS. How to get a PhD: a handbook for students and their supervisors. 2nd edn. Buckingham: Open University Press, 1994.

13. Writing by those whose first language is not English

I am not like a lady at the court of Versailles, who said: "What a dreadful pity that the bother at the tower of Babel should have got language all mixed up."

Voltaire (1694–1778)

Science knows no country.

Louis Pasteur (1822–1895)

Biomedical research workers whose first language is not English initially assume that this will reduce the chances of their papers being accepted by editors or conference organisers. They cannot believe that editors and conference organisers are only too delighted to be offered intelligible papers that contain original work with a clear message, that are scientifically reliable, of biomedical importance and suitable for the target journal or conference, whether general or specialist. Provided enough of these attributes can be identified editors will go to considerable trouble to help with the English (Whimster, 1991).

In the nature of things, many authors whose first language is not English have trained and may now work in difficult circumstances. Library facilities, computerised reference networks, books or journals, let alone modern word-processing or typewriting equipment, and even a regular supply of paper, all may be limited; and they may have laboriously taught themselves English from books or broadcasts, in isolation. Indeed, many whose first language *is* English also work in such circumstances, as I have myself, but usually with the advantage of being familiar with large international medical school or university surroundings. So the authors whose first language is not English may be unfamiliar with the planning needed (Chapter 8), particularly with regard to the target journal.

They may well, therefore, merely list the most prestigious general journals they can think of and submit their paper to the top one, without carefully complying with its Instructions to Authors or assessing it for its suitability for their work. If, in addition, the presentation is unattractive, with poor paper, poorly drawn figures and poor typing (or even handwritten), it is no surprise that it is rejected. After several rejections down the list of journals, such authors may be forgiven for believing that the editors are prejudiced against them, and think that it is because they are foreign, their names unknown and their institutions remote and obscure.

I believe that this is not so. Most editors are prejudiced primarily against work that is not original, not scientifically reliable, not particularly important biomedically or not in their field. They may judge these criteria themselves, but will usually have the originality and scientific reliability judged by referees (Chapter 14). If the work clearly fulfils the above criteria for acceptance, the authors must understand and not be upset by the likelihood that they will usually be asked to make some modifications. Any problems with poor typing and so on should be overcome by the journal office. Such problems may, however, influence the editor or conference organiser if the biomedical importance is marginal. The English can also be sorted out in the journal office if it is intelligible, but if the editor cannot understand what the author is saying, or if the message, and the proof that it is true, are not clear to the editor or the referees, the paper is likely to be rejected, just as any paper from any author may be.

If the paper is rejected, the referees usually go to considerable trouble to explain their reasons for advising against publication, and these are sent to the authors, but non-English workers may not know how to benefit from them, or not realise that it would be sensible to modify their papers accordingly before submitting them elsewhere. Not to take the referees' comments on board is a waste of their experience.

Very often the work is of great interest in the local context (Whimster, 1981) but it, or something like it, has been reported from elsewhere in the world already. The author often shows this lack of originality himself by referencing such reports. Nevertheless his work may have original points, differing from anything in the previous publications, which he must learn to bring out.

He must ask himself who will benefit from reading about the work. Do the local doctors need to know about particular diseases locally, for example the apparently high incidence of lymphoma in Bahrain (Whimster and Al-Hilli, 1986)? If so, why not present and/or publish it locally so that local practitioners will know about it? This will be biomedically acceptable provided that the local journal is listed in the databases, so that the paper is retrievable worldwide. The argument against this is that local journals have very small impact factors and carry no prestige for promotion and so on, which is undeniably true.

A solution is for the work to be published locally and then somehow to be transmitted to a wider audience later. This runs the danger of being condemned as "duplicate publication" by the International Committee of Medical Journal Editors (Appendix 1B), precisely because it might attract two doses of prestige for the same work. The danger is reduced if the local publication is in another language and if the second publication acknowledges the local publication.

So far, I have assumed that the author whose first language is not English is nevertheless competent in English. Many are perfectly fluent and confident in spoken English but are either not competent or not confident in written English; some believe that they are more competent than they are and others that they are less competent than they are. Of course there are many forms of spoken English throughout the world (McCrum *et al.*, 1986), but written English is less varied. However, just as it is almost impossible for foreign speakers to sound infallibly English to native English speakers, so it is difficult for foreign authors to develop the instincts to make their writings "sound" infallibly English to native English readers.

Most such authors worry about their English and believe that it will prejudice editors of English language journals against their papers. On the other hand I believe that editors will not be prejudiced against such a paper unless they cannot understand what the author means or unless the paper is already borderline in terms of its message and proof, originality, importance or suitability for the journal. The added effort required to make it sound English may be just too much for the editor to face.

By what means can foreign authors make their papers sound more English to improve acceptability for the borderline paper (Whimster, 1977)? There is no easy or infallible answer. While acting as English language supervisor for the English language journals published by the Finnish Medical Society Duodecim, I worked on papers that had already been accepted. I found that some Finnish authors wrote their articles in English and then got a native English speaker to make it "sound English" (Collan *et al.*, 1974). The success of this approach depends on the native English speaker. You may have to make do with whoever is available, but, on the whole, colleagues in the biomedical field were more successful than a long-time expatriate Englishman who was not in the biomedical field and not in touch with contemporary English. Teachers of English were particularly unsuccessful, because they tended to be pedantic about the grammar and not to know the words and expressions current in biomedical writing.

Other authors wrote their articles in Finnish and then either translated them themselves or employed translators to translate them. Those who did their own translation tended to have worked their thoughts out clearly in order to do so. Professional translators tended to be further out of touch with biomedical journal requirements and style than the authors themselves and were definitely not worth the high fees they charged. None of these strategies protects foreign authors, particularly if they have read much British or American biomedical "literature", from the language problems that native English writers have themselves.

Foreign authors often know that the spelling, use of words (for example"elevator" versus "lift") and style of writing in American journals differ somewhat from those in British journals (Appendix 6), but cannot be expected to work out how to use them to their advantage. How can they know when they, entertainingly, create words that do not exist, such as "fallacity", "unprecise", "evoluated", "aggressivity", "pathognomonocity", "progradiate", "neoplasmatic", "coarsely-granulous", "provosing", "debuting" and "hourse"?

How can they be at home with the associations that words acquire which make them inappropriate in certain circumstances? For example, "caseation" is not just tissue that has a cheese-like appearance; it should not be used if tuberculosis is not meant. "Denomination" and "ordination" have ecclesiastical overtones, so that "cirrhosis and chronic hepatitis were included in the same denomination" and "the ordination of wheel chairs" sound peculiar to English ears. "Roentgenological symptoms" and "findings occurred" sound illogical.

As a distance language supervisor I would prefer to go through the papers with the authors face to face. It would be so much easier to say "Just tell me what you want to say" and then distil what he tells you into contemporary English for him. Without being able to discuss it with the author, I do not know if any changes I suggest are changing his meaning rather than just making it clearer, and he may not know either.

In the end, authors whose first language is not English and who wish to have their English dealt with before they submit their paper should seek language supervision from the most suitable, sympathetic and biomedically related person available (Whimster, 1987,1989). It is a time-consuming business and therefore likely to be expensive. I hope the other chapters in Section C may also help these authors towards self sufficiency with the language.

References

Collan Y, Lock SP, Pyke DA, Whimster WF. Medical English for Finnish doctors. British Medical Journal 1974;i:627–9.

McCrum R, Cran W, MacNeil R. The story of English. London: BBC Publications, 1986.

Whimster WF. How can foreign authors help themselves? In: Lock SP. Thorne's Better Medical Writing. London: Pitman Medical, 1977: pp. 97–106.

Whimster WF. Self interest and local journals. Bahrain Medical Bulletin 1981;3:81–2.

Whimster WF. Colleague treatment. In: Sahni P, Pande GK, Smith J, Nundy S, eds. Better Medical Writing in India. New Delhi: National Medical Journal of India, 1987; 13–17.

Whimster WF. Getting the writing right. In: Naik SR, ed. Better Scientific Communication. Bombay, 1989: 53–7.

Whimster WF. Authors whose first language is not English - how should editors cope? European Science Editing 1991;42:7–8.

Whimster WF, Al-Hilli F. Lymphomas – are Bahraini Arabs really different? Journal of Pathology 1986;148:124A.

D
THE JOURNAL OFFICE

14. The editorial process

There's nothing so hard as minding your own business and an editor never has to do that.
Finley Peter Dunne

If you can visualise how the editor works you may be able to avoid mishandling or misunderstanding him. The life of an editor, especially an editor who is primarily a busy physician or scientist, is more complicated than you might think. Apart from computerisation, his responsibilities are still much as they were comprehensively described by O'Connor in 1978. You can imagine your precious paper and its covering letter in the journal office, waiting to be logged in and numbered for initial reading by an editor to decide what sort of a paper it is and to whom it should be sent to be refereed (peer reviewed). Yours may be the only one, or there may be many for him to consider, so he may be in inclusion or exclusion mode when he looks at it. As he does so, he is likely to consider the covering letter and whether all the required items are there (Chapter 11), to think about the paper's relevance and potential importance to his journal and its sphere of biomedical interest. So, more or less subconsciously, he will take in the title, the presentation, the length and the pictures; he will look for the message and he will read your covering letter. He will be affected by whether his instructions to authors have been followed and how much editorial work will be needed if he accepts it.

Acceptance or rejection

The editor may conclude that the paper is outside his journal's field or that it is libellous, blasphemous or totally irrational. Such papers will be rejected without further consideration (fig. 14.1). He will reject others because they are obviously not original; obviously trivial; without a message or with an unsubstantiated message; so unattractively presented and/or muddled that it will be too laborious for him or his staff to get the author to rewrite it to an acceptable standard or to do it themselves; or because he knows enough about the partic-

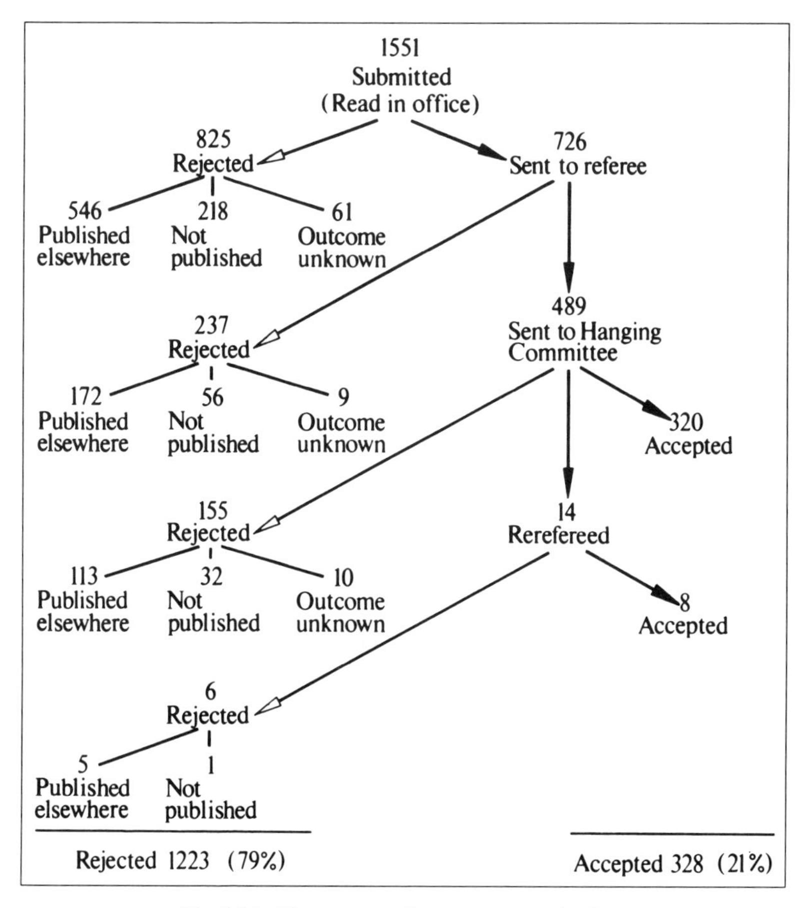

Fig. 14.1. The process of acceptance or rejection.

ular field to trust his own judgement that it is not sound. In journals with full-time editors, even papers due for early rejection will usually be read by a second editor. The rejection letter will usually be a definite rejection, against which there is no point in appealing.

He may decide that he has a winner on his hands, a paper of such importance that it must be given to his readers with the minimum delay, along the lines, for example, of Watson and Crick's paper on the structure of DNA (1953). Editors long for these but have to check and double check their first impressions to ensure that a winner really is a winner and is not flawed in some way that will later be embarrassing.

For the rest, the editor will seek the advice of one or more referees (Chapter 15), chosen from the journal's database for their area of interest and expertise, their approach to refereeing, their speed of response and possibly in accordance with the author's suggestions in his covering letter.

On receipt of the referees' advice the editor may reject or accept the paper outright, seek advice from another referee (usually with reference to the statistics), accept it subject to specific modifications, suggest modifications and resubmission, or have another think before deciding. The latter course applies particularly when there are too many run-of-the-mill but acceptable papers for all to be published. Some of the big general journals have a committee to decide which of these should be accepted. It is surprising how varied the opinions of the committee members can be – one arguing strongly for and another strongly against the same paper – illustrating how subjective and indeed complex the paper writing/publishing process is. Authors have to accept this, as there is nothing they can do about it.

Editorial process time

Authors are, however, concerned about how long it will take for a paper to be accepted or rejected. Many journals state in their Instructions to Authors how long they expect to take, but may well take longer if the decision is difficult. Authors can expect an acknowledgement of receipt of the paper within a few days and can reasonably contact the editorial office for progress if they have had no acceptance or rejection within, say, two months. Telephone calls are not welcomed for obvious reasons, but polite enquiries by fax will usually produce a speedy reply. If the delay becomes too long the author is free to withdraw the paper and submit it elsewhere.

If modifications are suggested, either to the paper or to the work, usually on the basis of the referees' reports with which, these days, the author will be supplied, the author has to decide whether to argue about all or any of them in his reply to the editor, make them, or withdraw the paper and submit it elsewhere. Some suggestions result from misunderstanding by a referee, who may have misunderstood because your writing is not clear, not necessarily because he is ignorant or biased! Others suggest that the work is incomplete and that more is required; this may be impracticable. Having decided what he thinks about the referees' suggestions, the author should write back to the editor responding in sequence to each suggestion, usually enclosing a further set of copies of his paper. If he has incorporated new material into his script, it is most helpful to the editor if the changes are made easily recognisable by highlighting or some other device. The new version has to have the signatures of approval of all the authors, just as the initial submission had. Authors must be allowed to dissociate them-

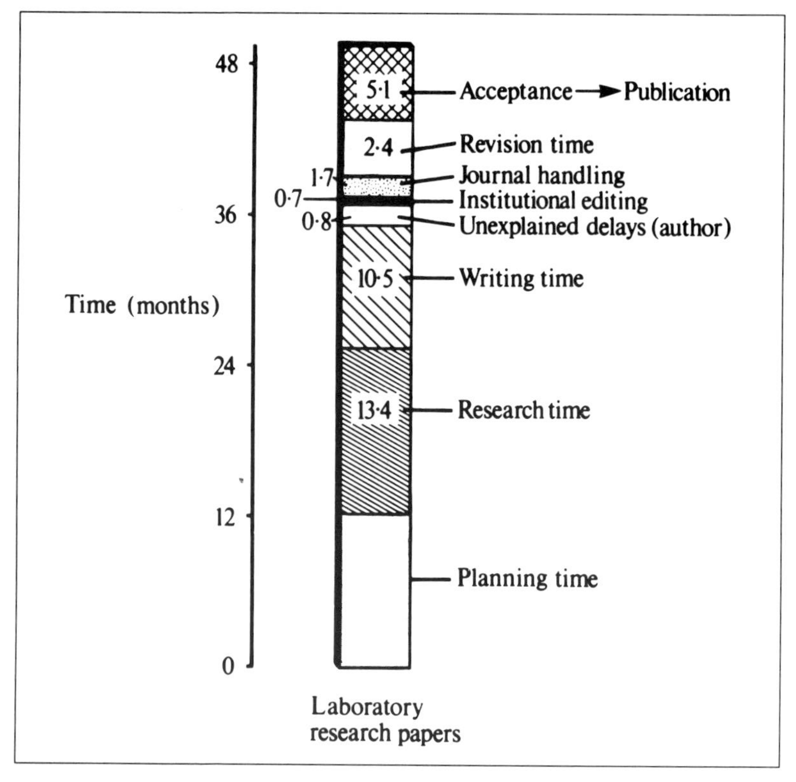

Fig. 14.2. The time lapse between hypothesis and publication.

selves from the work as now presented. It is interesting that the greatest delay in the publication process is the time it takes authors to respond to editor's requests for modification (fig. 14.2; Roland and Kirkpatrick, 1975; Lock, 1985).

The editor then has to decide whether he accepts or rejects any or all of the author's changes or arguments, or whether the referees should see the paper again, in which case the process is repeated.

Fortunately the editorial/refereeing process tends to refine the paper. Even if it is ultimately rejected, it will be usually be in a better condition to be accepted by another journal. In Lock's (1985) series (fig. 14.1), 836 (73%) of rejected papers with a known outcome were subsequently published elsewhere, 80% of them without change. Many went to specialist journals, which may have been the recommendation of the referees, but nevertheless ignoring and thus losing the benefit of much of the work of the referees. Perhaps authors should take comfort from that, if not the readers, referees or editors.

Subediting

Once the paper has been accepted, the editorial staff will subedit it (this process is also called technical editing or copy editing). Authors will see what this involves from the Checklist for Technical Editors in Appendix 2. Excellent practical advice if authors ever have to do it themselves is available from O'Connor (1986), or in a slightly more jocular approach from Morgan (1986) who argues that editing is the least understood medical speciality. In practice the amount of subediting a paper gets varies considerably. It depends not only on the quality of the paper but also on the resources of the journal. Obviously less will be done by journals with a part-time editor, part-time subeditors/secretarial staff and remote editorial board members – the manpower resources of many specialist journals.

Journals that use language supervisors (Chapter 13) send them the paper at this stage. Language supervisors may also comment on the length, layout and logic of the paper, and thus also act in the interesting role of additional, post-acceptance, referees or consultants. A dialogue with the author about meaning may have to be conducted, which can be fruitful, but irritating to the author who believes that publication is being delayed. He can be reassured that this will not be the main delay (fig. 14.2).

Meanwhile, the subeditor will also be checking the grammar, spelling and punctuation, and internal consistency, that is, that the numbers in the text, figures and tables all correspond, and also that the references in the text and in the reference list correspond. The corresponding author may also have to be contacted by the subeditor to clarify inconsistencies. Changes to make the paper correspond to the journal's "house style", its policy on nomenclature, spelling and units of measurement, will be made. The layout of tables and figures will be determined. The paper will be "marked up" for the printers, with type sizes and typefaces. Some journals do a certain amount of rewriting to make the papers clearer, especially if the author's first language is obviously not English (Chapter 13), but subeditors and language supervisors have to try not to change the author's meaning, although neither they nor the foreign author may recognise that it has been changed. I can see no solution to this dilemma.

References

Lock S. A difficult balance. London: Nuffield Provincial Hospitals Trust, 1985.

Morgan P. An insider's guide for medical authors and editors. Philadelphia: ISI Press, 1986.

O'Connor M. Editing scientific books and journals. Tunbridge Wells: Pitman Medical, 1978.

O'Connor M. How to copyedit scientific books and journals. Philadelphia: ISI Press, 1986.
Roland CG, Kirkpatrick RA. Time lapse between hypothesis and publication in the medical sciences. New England Journal of Medicine 1975;296: 1258–64.
Watson JD, Crick FH. Molecular structure of nucleic acids: a structure for deoxyribose nucleic acid. Nature 1953;171:737–8.

15. Peer review

This is the interpretation of the thing... Thou art weighed in the balances, and art found wanting.
Book of Daniel, chapter V, verse 27

Although editors have papers peer reviewed or refereed for their own purposes, authors can benefit greatly from studying the guidelines offered by the target journal to its referees (Appendix 2), finding out what the referees are asked to do by the journal, and studying the referees' reports that are nowadays sent to the authors. To be refereed is a character-forming experience for an author, which certainly improves his publishability, although many authors do not take advantage of it, and apparently do not have to (see Lock, 1985, and Chapter 14).

For practical purposes "to peer review", i.e. to review one's peers or equals, is synonymous with "to referee". The terminology is, however, thoroughly discussed by Lock (1985), who shows that both terms have certain differing overtones of meaning. Reviewing has a longer history than is perhaps generally appreciated and was regarded as desirable from the start of scientific journals with the *Philosophical Transactions of the Royal Society* in 1664. The Council of the Royal Society ordered that the Transactions should be printed "being first reviewed by members of the same [Council]". Soon, too, the members of the editorial board of the *Journal de Scavans* in Paris were meeting regularly to discuss articles for publication. Today's authors might consider Jenner's state of mind when, two hundred years ago in 1796, this refereeing process lead to the Society rejecting his account of the first use of vaccination against smallpox.

One or more referees

In practice, if, as is now usual, the editor wants advice about the paper, he sends it to one or more referees for their opinion. Until recently there was little guidance for referees beyond baldly asking them to answer the key questions of whether the paper was original, scientifically reliable, of clinical (or biomedical) importance, or suitable for the journal (Smith, 1982). One could reply "yes", "yes", "no", "no", without being very helpful. So each referee concocted his own form of constructive or destructive advice around those four points. Clearly referees need advice, too. They often need reminding that they are advising the editor about those four points, not instructing him to accept or reject the paper. Some journals produce more extensive questionnaires, which, like most questionnaires, are difficult to complete satisfactorily because the questions do not quite hit the bullseye. Others publish guidelines (Appendix 4). Most ask the referee for confidential advice on acceptability, both generally and for that particular journal, plus a report that can be sent to the author. This saves the editor from distilling suggested modifications and the reasons for them from a referee's report himself, and puts the task into the probably more specialised hands of the referee. It also stimulates the referee's sense of responsibility to produce a well argued analysis for any suggestions for modification, while the actual advice on the four points is presented confidentially, as mentioned above. Many referees spend many hours producing a constructive critique for the authors, contrary to many authors' belief that referees are negatively critical, delay publication, perhaps for their own ends, and plagiarise the author's work.

Statistical referees

A useful innovation is the introduction of statistical referees, without whom the support for many messages would be inadequately assessed. Many initial referees are capable of detecting flaws in the statistics, so editors may reserve their statistical referees for secondary refereeing which they find necessary after the first referees have submitted their reports. Some journals publish guidelines for statistical referees (see Appendix 2). This may prolong the refereeing stage but is essential if untrue messages are not to be published.

Referees for ethical, commercial, fraudulent or other issues

With the increasing concern of the public for the ethics of medicine, and indeed their willingness to sue if any medical or scientific outcome is undesirable, journals will certainly come under fire, under the provisions of the Declaration of Helsinki (Appendix 3), if they publish any unethical work. Until now, whether the work in an accepted paper is ethical has been judged by the editor (and perhaps commented on by the referees, one imagines, although there are no instructions for them to do so). An independent referee's opinion on the ethics, for example of a paper on child abuse, may become an essential safeguard for the journal. Similarly the sale of the products of a commercial organisation could be affected by a paper in a journal. This has been addressed by the use of structured abstracts (Chapter 9) and within some journals' Instructions to Authors. More may need to be done and litigation may be averted if an independent referee has judged that the paper is a fair representation of the findings. An expert referee may also be needed to ensure that a paper is not fraudulent, the result of plagiarism, or pirated. On the other hand, many authors worry that the referees themselves may take advantage of their privileged position of seeing the work before publication to gain precedence or prevent or delay publication. The state of play in this area has been set out by Lock and Wells (1993).

How to referee

There must be many ways of refereeing a paper. The easiest and quickest approach is to say that the paper is original, scientifically reliable, of biomedical importance and suitable for the journal. The most time consuming is to criticise every small detail – as well as doing the editing, which is no part of the referee's task.

I myself, however, prepare my report directly into my word-processor as I read, by summarising what the authors set out to do, what they did, what they found, indeed distilling the paper as I go through it, trying to answer from the text the questions that arise in my mind. In this way I soon know what they thought their message was and whether I can believe it from the evidence presented, and, more important in the referee role, if I cannot, why I cannot. The commonest difficulties are an ill-thought-out message; not being able to understand what was actually done, and hence being unable to know whether it supports the message; not being able to understand what was actually found, together with impenetrably presented statistics; and, particularly in case reports, lack of originality, very often revealed not by my own knowledge but by the references given by the authors themselves, with no attempt to find original features that would justify publication.

A further advantage of this approach is that, when the editor sends me his decision together with any other referees' reports (an admirable recent innovation), I can easily compare what I said with them. I find that the points we have commented on usually differ markedly, even if the conclusions with regard to acceptance are frequently the same. Furthermore it is of value when the editor returns the paper to me for re-refereeing with the author's comments and modifications.

Fairness of refereeing

There has been some discussion about the fairness of refereeing of biomedical papers and whether naming or "blinding" of referees would make refereeing fairer (Lock, 1985). With regard to anonymity, I try to be as objective and constructive as possible as a referee, which takes quite enough time and energy as it is. I am not myself prepared to enter into a dialogue with the author; I am advising the editor. Therefore I do not wish to be identified and would not continue to referee if I could not remain anonymous. Perhaps referees' wishes should be ascertained by the editor?

With regard to removing all evidence of authorship from the paper, i.e. blinding the referee to its origins, whether humble or prestigious, I believe that that is impracticable, messy and, I hope, unnecessary for me. I am aware of the belief of some overseas authors that Western journals and their referees are prejudiced against papers from developing countries. In my experience, editors and referees go to considerable trouble to help authors from other countries and other cultures to publish work with an interesting message and satisfactory supporting evidence, particularly when the author's first language is not English (Chapter 13). Unfortunately, work reported in many of these papers is not original and/or satisfactory, sometimes because of a lack of understanding of how Western biomedical research and publishing works. Also, the authors may not appreciate how many papers from Western countries are rejected for the same reasons. I hope that this book might particularly help them all.

Finally, referees may need protection. With the increasing demand worldwide on biomedical staff, both hospital and academic, to show their productivity by the quantity of their publications, and since the quality, importance and future applicability and value of the work are so hard to judge, there are many more papers to be refereed. Most authors would, I believe, not wish the standard of refereeing, whatever that is, to decline. Most referees would not want more papers to referee. So more referees must be recruited, and trained, hopefully in the precepts set out in this book, until perhaps every author/practising biomedical researcher does his share of refereeing. Readers of this book are obvious potential recruits. Would this be better than having "professional" referees, i.e. paid, full-time, and soon out of touch with the field?

In this respect, perhaps the Internet could be a threat, in that authors could merely put their papers on the Internet and bypass the whole journal process, both editorial and peer review. On the other hand, unedited, unrefereed papers could be denied the rewards of publication. Authors could be required to name their referees. Refereeing also attracts "CME" (continuing medical education) points, so the politics proliferates.

References

Lock S. A difficult balance. Editorial peer review in medicine. London: The Nuffield Provincial Hospitals Trust, 1985.

Lock S. Wells F. Fraud and misconduct in medical research. London: British Medical Journal, 1993.

Smith R. Steaming up windows and refereeing medical papers. British Medical Journal 1982;285:1259–61.

16. Dealing with acceptance and rejection

If you can meet with Triumph and Disaster and treat those two impostors just the same...
Rudyard Kipling (1865–1936)

Acceptance

The letter of acceptance finally comes. You and your co-authors congratulate each other and get on with the next project. Some weeks or even months later another envelope hits your desk: printer's proofs of your paper, saying "Please read, answer any 'Editor to Author' questions, make any corrections and return within 48 hours". This means that the editor has sent your paper to the printers and that it has a place in a future number of the journal. Some journals very helpfully enclose your manuscript with the subeditor's marks on it so you can see what has been done to it.

What to do with the proofs

You may notice that some corrections have already been made, usually in red. These have been made by the printers themselves and already taken on board. There may be some questions in the margin marked "Ed to Au", meaning "editor to author", which you must answer. Do not make your own corrections in red. Take a pencil (so you can rub out any marks you decide to change), the marked copy, if you have been sent it, or your file copy of the paper, and with these beside you start with the proof title and methodically read through *everything* in the proofs to:

- ensure that your paper has been completely accurately transcribed into print
- ensure that you did not made any silly mistakes in your paper
- check each reference in the text with your originals and the reference list.
- recheck the details of each reference with your photocopies
- check each illustration, table and figure for its caption and every detail on it
- recheck to be sure that the totals add up correctly

You will always find something to correct. If you have found nothing, you have missed something, so look again.

Mark clearly any error you find in such a way that the printer can change it to what you want, using plain English if you wish, but, more easily for him, using accepted proofreading marks. You may not be familiar with these but the common ones are set out in Appendix 8 with an example. You may not like some of the rewording, but this is the editor's choice so consider it very carefully before you change it back to your original – the editor's version may be clearer or have some other virtue. You will not be popular with the editor if you try to use the proofreading stage as an opportunity to rewrite any of the paper. If you have additional new data you can try to negotiate with the editor to include them in an addendum. There is no point in correcting spelling, capitals or hyphens because they will have been subedited into the journal's house style (Chapter 14). Return the proof in the addressed envelope provided.

I always keep a photocopy of the corrected proofs so that, if they get lost in transit, I can see what answers and corrections I made and can more easily do the same again. It is also nice to have it in the file to show the co-authors, and to refer to legitimately as "in press" in the next papers. Before it has been accepted the paper has to be referred to as "in preparation", which cuts little ice with anyone, since it may never come to fruition.

Also you must complete the order form for offprints (extra copies of the article made at the same time as the journal is printed) or reprints (copies of the article reprinted separately afterwards, often with a different layout or with a cover). The author whose address has been given for correspondence (usually you if you have been sent the proofs) will get the reprint requests, but you should distribute some reprints for your co-authors' collections. Send the form back – to the right address, which may be different from the one the proofs go back to – enclosing any necessary payment.

Await the arrival of the journal itself with restrained delight – you will always find some error that nobody spotted. Something as obvious as an x-ray back to front? Never mind, no reader that you meet will have spotted it.

Rejection

Many authors' papers are returned to them with a rejection letter plus, usually today, the referees' reports. This is disappointing and humiliating, but may merely mean that you chose the wrong target journal or did not fulfil its requirements. It may in fact merely reflect, in these days of increasing pressure to publish in high-impact-factor journals, the fact that many such journals now receive many more papers than they can possibly publish, so even publishable ones have to be rejected. If this is the case, it is to be hoped that your target journal is honest enough to say so.

After reading the rejection letter and referees' reports very carefully, you need to determine whether to abandon your attempt to publish your paper in this journal, if the message has now been superseded, for example; to appeal to your target journal; to start the submission process again with a new journal, perhaps in another form, for example as a short report (in a journal which accepts this format) or as a letter; or, if really desperate, to publish it as a monograph, or start your own journal. A well written letter can be a useful and rapid form of communication, especially for reporting isolated clinical or scientific observations which do not need lengthy discussion or documentation, and those that contain new or unusual facts may be indexed in the databases. It should be sent to the editor marked "for publication" and consist of not more than 250 words and two or three references.

An appeal may still be the most painless way to achieve publication, succeeding in perhaps 15% of cases. It is useful to have had a civilised and good natured correspondence with the editor up to this point. The grounds to consider putting forward for a review of the decision are that:

- the journal really is interested in this subject, as shown by previous publications in it, including perhaps your own
- the arguments put forward by the referees are not sound, as shown by your counter-arguments
- the message is more important than the editor realises
- the evidence is all that anyone is going to be able to collect
- further publications (quoted) add to the value of yours

I have to admit that in the case of a further paper in a series of publications in the American Review of Respiratory Disease, which was rejected by a new editor, my appeal failed. "C'est la vie" was the only possible reaction. Nevertheless the appeal process can work. For an insight into peer review at work, read the fascinating dialogue between author, referees and editor in the *British Medical Journal* (1985).

Submission to another journal, perhaps a more specialised one, is the next option. A pause of a few weeks to clear your mind and think around the subject is helpful. Has anything new been published on this subject? Then you should take the opportunity to replan your targeting strategy by studying the Instructions to Authors of other journals, the journals themselves and their impact factors, plus the new insights you have received from the rejecting journal. In particular, rethink your message, and the evidence you are presenting to support its validity. Be clear about the answers to: Why did you start? What did you do? What did you find? Rethink: What does it mean? You may find that considerable rewriting is demanded – but some of the old writing still in your word-processor will certainly be usable.

Reference

BMJ. Peer review at work. British Medical Journal 1985;290:1555–61.

E

PRESENTING RESEARCH AT MEETINGS

17. Planning presentations

Reading maketh a full man; ***conference a ready man;*** *and writing an exact man.*
Francis Bacon (1561–1626)

Newcomers may not at first understand that workshops, meetings, seminars, symposia, conferences and congresses are where the members of the "invisible college" of their speciality or field of interest meet to interact. On the positive side, this means exchanging news of how everyone's research is getting on, what results are emerging and where they agree or conflict, where collaborations might be fruitful, who is planning to do what, and where there might be a job coming up. On the negative side, you learn where and for whom the research, the job or life in general have gone wrong, and who is at loggerheads with whom. You must decide whether you want to become an insider or whether you have what it takes to be an outsider in the short or long term, perhaps until retirement.

It does no harm to attend the business meetings that are often held during conferences. If you are not a member of the society whose business meeting it is, you can usually obtain permission from one of the officers to observe but not to vote. You will learn who is who in the society, how the officers handle the meeting and the finances, who the "godfathers" are, who are the up-and-coming young bloods, whether the society would be an agreeable one to join, what rival societies there are, what future meetings are in prospect and what plans the society has for development. This may be very important for your own research development.

This is the context within which you must be clear why you want to present your work in public. It is probably because you have completed your project and obtained some interesting results: you want to try out what you think your message is on a knowledgeable, critical, but not too unsympathetic audience; to obtain their reactions before publishing your work in a high impact journal and advancing your career. On the other hand your boss, with good or ill will in his heart, may have told you to present it for his own reasons. In any case it is valuable, instant, public peer review which you can use to improve the work.

Having analysed why you want to do it you must target your forum. Do you want to engage a general audience within your speciality or a specialist audience? In big conferences there will be simultaneous sessions, so the people whose reactions you want to hear may be elsewhere at your time. Workshops may be too specialised. Regular local, regional or possibly national meetings of your speciality, especially if it is a small one, are likely to welcome newcomers and give them a fair hearing. In the UK the Royal Society of Medicine has many speciality sections attended by a wide variety of enthusiasts. Many of the sections also hold meetings for medical speciality trainees to present their work, as do the Royal Colleges.

Does the target meeting publish the abstracts or the proceedings? As far as prestige goes, you should be aware that, although published abstracts can be included on your curriculum vitae, they carry little weight unless followed by publication in the proceedings of the meeting, which may be useful to establish precedence, or as a separate publication. Meeting proceedings, often arranged and sometimes paid for by the organisers to appear in a journal or journal supplement, carry less prestige, even if the papers are independently refereed, than a definitive publication in a peer reviewed journal. You also have to be careful that you do not run into conflicts of interest if your work has been supported by a commercial organisation, such as a drug company on whose product you have been conducting trials (Appendix 1B).

Getting on to the programme

Start by studying the successive announcements, which usually include calls for registration and abstracts, as well as stating when and where it will take place, what registration, travel and subsistence will cost, what you must do to get on the programme and by when, the scope of the meeting and who is on the scientific committee. The latter gives you an idea of its standing and who will accept or reject your application. You will probably have to submit an "abstract" to the committee (Chapter 18) and say whether you wish to present a paper (i.e. orally, from the podium, with yourself speaking, actor-like, before the assembled multitude) or a poster. Even if you say you wish to present your work as a paper, the committee may accept the work only as a poster, to which you may agree or not. In general, and wrongly, I believe, poster presentations are regarded as having less prestige.

Conference organisers often seem unclear about what criteria to use to decide whether an abstract should be accepted for oral presentation or as a poster. Complicated questions with complicated, perhaps controversial, results that need explanation and audience discussion are best presented orally, since the author does have the audience's attention for perhaps 10 minutes. The poster format is best for presenting a clear question with clear results leading to a clear answer (including negative answers), since the individual viewer will probably devote no more than two minutes to viewing it, as there are often a hundred or more to get through. Unfortunately, both organisers and participants think that the "best" work should be presented orally.

Nevertheless, authors should use these criteria to submit suitable abstracts specifically for poster presentation, without feeling that they are second-class presentations, except, of course, that the glory of the personal appearance must be foregone. Organisers should know what they are doing when they offer authors the option of converting surplus oral presentation abstracts to posters, especially if time is short, since posters can be time consuming and costly to prepare.

The planning sequence now includes preparation of the abstract for submission by the closing date for abstracts, submitting it in the required format, being accepted, concluding the work and preparing the presentation, as described in Chapters 18–20.

18. "Conference Abstracts"

They are abstracts and brief chronicles of the time: after your death you were better to have a bad epitaph than their ill report while you live.
Hamlet (Act II, scene ii)

In recent times, "conference abstracts" have fallen into disrepute because some conference organisers have published them whether they were refereed or not, whether they were accepted for the conference or not, and, if accepted, whether the authors had turned up and presented them or not. It may also be impossible to change the abstract before it enters the public domain (under your name) if other participants show your ideas or interpretations to be wrong. Authors have, however, happily included them on their curricula vitae.

The problem is that conference organisers have confused the requirements of journal paper abstracts (Chapter 9) with those of conference abstracts (sometimes laying down that they should include the materials and methods, results, conclusions and significance (Waxman and Dudley, 1983; Gourgoulianis *et al.*, 1985). In fact the essential purpose of the conference abstract is to attract and gain the acceptance of the conference organiser. Perhaps it should be distinguished from the paper abstract by calling it the "conference preview" (Whimster, 1988) or some other term.

The conference organiser needs a clear title and some indication of the message and whether it is to be believed, but not to reveal all, like a good striptease artiste leaving something good and unexpected for the performance. One hopes that if it succeeds in attracting the conference organiser it will also interest the conference participants.

You must also follow the conference instructions, such as fitting the abstract into the camera-ready frame on the abstract form provided. This form often has pale outlines so that they will not appear when printed in the abstract book. Camera-ready copy reduces the cost of printing the abstract book by eliminating typesetting, although some typesetting is usually necessary because registrants do not follow the instructions, and, for example, send photocopies of the form. It is helpful to put the name of the author whom you expect to present the paper first and, in case anyone might want to get in touch with you after the conference, include your address, phone, fax and e-mail numbers.

You may find it tricky to word the text, since the conference may be many months ahead and you may not know what data you will have by then or whether the findings will be positive or negative. Give just enough to make your presentation sound novel and interesting, emphasising any new techniques or interesting uses of old ones. Give some data if you can and avoid the off-putting expression "...will be discussed". You have to get your abstract into the box, but you are not obliged to fill the box. Provided they are intelligible, participants welcome short pieces to read.

It is helpful to the participants to include any key references in the abstract, so that they will be readily available in the abstract book, which often turns out to be the most lasting and valuable product of the meeting.

The abstract book is where the conference abstracts should be published; it is not strictly in the public domain, although an ISBN number can be allocated to the abstract book, which makes it retrievable.[1] It is also helpful if the abstract book has an alphabetical index of contributing authors with as much information (address, phone, fax, e-mail numbers) about them as possible, since this one way of being able to locate them and their work afterwards.

1 ISBN numbers are allocated by a responsible agency in each country. In the UK they are issued to publishers by the Standard Book Numbering Agency Ltd. (part of the Whitaker Group).

References

Gourgoulianis K, Panagakis A, Tsakraklides V. Anatomy of the IXth European Congress of Pathology Abstracts. Pathology Research and Practice 1985;180:246–8.

Waxman BP, Dudley HAF. A critical assessment of the submitted abstracts for the 1982 Winter Meeting of the Surgical Research Society. British Journal of Surgery 1983;70:182.

Whimster WF. Conference previews. British Medical Journal 1988;297:1000.

19. Presenting papers at meetings

No one wants to hear a speaker talking like a book.

Clifford Hawkins (1915–1991)

There are as many ways of preparing a presentation for a meeting as there are speakers, although to judge from the "How to do it" books, you would think that their writers' was the only way. One reason for the variety is that the medical specialities and biomedical research areas themselves are very varied, from epidemiological work with lots of statistics, through sociological, psychological, caring and clinical fields, including clinical trials, to laboratory work and "basic" science, some much more pictorial than others. Nevertheless, telling the audience what you looked for and what you found has to be done, within the time allotted, by the two components of speaking and illustrating, i.e. reaching the audience's critical faculties via their ears and their eyes. You can also plan to elicit "audience participation".

The illustrations consist of slides, OHPs, film or video, and any other visual aids that you may think of, including patients. Listeners' and speakers' views on visual displays in oral presentations vary greatly, from those, like myself, who love plenty of ingenious visual displays, to those who find them distracting, especially if they are shown for too short a time.

Design

I start by designing my talk, working out my message(s) and what supporting evidence I need, and my conclusions, in list form on A4 ruled feint paper with plenty of lines. I take into account that a paper for speaking differs from a paper for publication.

	Publication	Speaking
Introduction	10%	30%
Methods and Results	50%	40%
Discussion	35%	20%
Conclusion	5%	10%

This rough division of time shows that in oral presentations more time must be spent on switching the audience into the problem and on what your answer means, with less emphasis on actual methods and detailed results. These may be explored in the audience discussion afterwards. There may have to be some repetition and padding to allow the audience to absorb the ideas. It does not include references, which should be in the conference abstract (Chapter 18). Acknowledgements can be put on a final slide.

I then list all the visuals (slides and/or OHPs) I would like to have to illustrate my show (Chapter 7). I do not worry about the number of visuals at this stage because I cannot tell how long I will spend on each one. I do know that audiences take in pictures much more quickly than words or graphs or tables, especially if I have an effective pointer. I know that double or triple projection uses many more slides and may go wrong. I give particular thought to, and usually write out, my opening and closing paragraphs in full, refining them over the days before the talk. I then design my visual displays, ready to be made into slides or OHPs, and rewrite my list with the opening paragraph, followed by the titles of the slides/OHPs, concluding with my closing paragraph.

Getting the visuals together

It is not a good idea to combine slides and OHPs in one presentation unless you know the venue and the equipment well, because the audience is distracted by the changeovers. For conferences, slides work best. OHPs work well and are more adaptable for small groups and workshops. I already have many slides and overhead transparencies from previous presentations, so I review those that may contribute to the new one. It is surprising how often the ones you already have are not quite appropriate for the new presentation, so you must take the trouble to make new ones, including at least a title slide. Audiences are not flattered by slides that were made for some other purpose, especially if they are obviously old, unless they are of something unique.

As to content, the books say that you should have only one idea on a slide. I don't know what they think constitutes one idea. I use *word* slides primarily to list headings of my presentation, so that the audience has a visual framework within which to organise what they are hearing and seeing on my *data* slides, tables and graphs, and my *picture* slides. The message should be firmly stated in *one* last slide. "Conclusions" that go on for several slides disturb the audience, who forget the earlier ones, and they give the impression that the author has not got a proper grasp of the work.

The same principles can be used to make a set of OHPs. The principles of making slides and OHPs legible, the type size and audience distance and so forth are set out in Chapter 7. The overhead projector can now have a transparent screen on a lead from a laptop computer put on it to show visuals from the computer, in effect overhead projecting the computer screen in black and white or colour. The equipment costs £2–3000. As it seems to be difficult to transmit enough light to give bright images, this works best with small groups, as do OHPs generally.

Rehearsal

Especially with a short proffered paper containing a new message and new data, for which 10 minutes are allowed with only 2–5 minutes for discussion, I practise it first on my own with the slides/OHPs to ensure that I can say what I want to say in the time. One can be rather more relaxed about longer talks. Next, I invite an audience to hear and comment on it. My audience is the usual group of colleagues within the department, who are used to doing this together and to which new people come and go. Then I go through it again on my own with their recommended modifications until I can do it in the time, with only the slides/OHPs as prompts. Each of these stages may involve rejecting some slides. I go through it a final time the night before the presentation, however late it is, and then put it into my hand baggage until the presentation. It is asking for trouble not to carry your presentation (notes and slides/OHPs) as hand baggage. Hold baggage might be left at home or in the car, or be despatched to the other side of the world. I have found that folding "slide wallets" holding 50 slides in rows of five are the most compact way to carry notes and slides. On one occasion, however, I failed to prevent a "helper" loading the slides into the slide carrier; he took them from right to left down the rows instead of left to right. It took half an hour to diagnose the problem and reload them.

On the day

I have found that, if I try to run through the presentation on the day, the adrenalin just muddles the sequence up in my mind. I am usually nervous as I approach the podium: *Will the talk work? Have I forgotten anything? Is there a horrible mistake in the argument?* On the rare occasions when I have not been nervous, my colleagues have commented on the lifelessness of the presentation. So adrenalin may be uncomfortable but it livens up the show. This may be hard to maintain if you are giving more than one presentation at the conference.

There are 15 important points to check before you start:

- check you have your notes and slides
- reread the organisers' instructions
- check when and where you must deliver your slides
- check when and where you must make your presentation
- check how long you can talk for and how long there is for discussion
- inspect the hall (preferably both when someone else is speaking and when it is empty)
- inspect the AV equipment (who will change the slides and dim the lights?)
- check if there is a microphone
- check that a pointer is provided and/or that you have your own
- insist on loading your slides yourself; take the opportunity to run through them
- await your turn at the front where the chairman can see you
- do not allow yourself to be the chairman for your own session
- STAND UP, SPEAK UP and SHUT UP (as old time preachers used to say)
- if anything goes wrong, such as a power failure, wait calmly for the chairman to sort it out, and decide whether to start again from the beginning
- remember to collect your slides afterwards

I keep my notes in my hand, out of sight. I do not expect to have to read them but the opening paragraph is there to get me going if I freeze. Similarly the final paragraph is there to get me off the podium if I forget the climax. I have never used them yet. I may look at my list of slides during my talk. If the slides turn out to have spelling mistakes, or are in the wrong order or out of focus, or anything else goes wrong, especially if it is obviously not your fault, try not to be apologetic or draw any more attention to it. Sail calmly on to the next point.

Speakers who read their scripts out, or indeed read directly off their slides, are rightly regarded as the bores of the conference. Reading written English aloud ruins the speaking rhythms, and rapidly becomes unintelligible. [It is just possible (I have done it when I have been unable to rehearse) to write a script for yourself in colloquial spoken English and read it concealed behind a lectern without anybody realising that it is being read. This is too risky to attempt except on your home ground.]

I appreciate a firm chairman; he has to be brave and forceful enough to stop speakers who overrun. One of the most difficult tasks for the speaker is to keep track of his time, especially if he did not start exactly as the programme promised. For this reason it is most unwise to be persuaded to chair yourself. I try to be firm with myself in asking the audience if they can hear and see, not to speak too quickly or too softly, and to use the microphone if so advised. You may have to allow time for translators to do their work, which sometimes brings disconcerting laughter at a joke you have moved on from. One is advised to stand still and not to fidget. In fact I find that I prefer speakers who move about freely, focusing on and involving different sections of the audience. I am not put off by non-specific audience reactions, such as going to sleep, since many people, including myself, go to sleep by reflex if the lights are dimmed for slides, and many may have jet lag. Some are old and tired, usually the ones you really want to impress. Complete darkness will offend compulsive note takers.

Having two speakers on the podium giving two halves of a presentation is surprisingly stimulating, since, if they interact, the participants seems to feel that they can participate.

Lecturing

Although Hawkins (1985) took rather a different view of the lecture, I do not think of presenting papers at meetings, even the "guest" or "keynote" lectures, as lecturing, which I reserve for teaching classes of undergraduates or postgraduates. These are rather different, often more cohesive, groups, whom you may get to know better as an audience, and usually speak to for rather longer, though hopefully not longer than 45 minutes. My listening attention span and my patience is much shorter than that. The principles of lecturing are, however, much the same as those set out for presentations. Many people believe that the lecture is a hopelessly inefficient form of teaching. I believe it is an extremely efficient way for one organised teacher to put light and shade into a chunk of a subject for the whole year of students (about 100) in half the time that each student could scan the flat acres in the library on his own, if he ever got round to it. It is far less labour-intensive than small-group teaching or tutorials, and can set the scene for those activities.

Handouts

Handouts should not be needed at meetings if there is a proper abstract book (Chapter 18). I do not like preparing handouts for lectures, except simple outlines, because as a listener I have always preferred to take my own notes. I have never understood why some lecturers prevent students from taking notes. I certainly have to integrate what I have heard by thinking about it afterwards in conjunction with my notes, not someone else's handouts. I should, however, be very happy for participants to be able to rerun videotapes of my presentations (see below).

Videorecording

One disadvantage of oral presentations is that those of the most charismatic and influential speakers are lost. They disappear into thin air and the audience is left with only the small percentage that they can remember, even when the spoken word is reinforced visually. There is nothing to pore over or analyse. Audiorecordings have long been made, but they are a very inaccessible way into the material. I have, however, used them to transcribe discussions for publication in the proceedings and found much more in them than I heard at the time.

It has been cumbersome to videorecord presentations until recently, but with light, small, hand-held camcorders with a "steady state" facility to eliminate camera shake, videorecording is within the reach of everyone. The trick is to ensure, by using the zoom, that everything on the screen is legible in the viewfinder. Do not worry about the sound; if you can hear the speaker the camcorder microphone will be sensitive enough to pick up what is said, but do not record next to the projector because the camcorder microphone will also pick up the noise of that. Do not worry about being an amateur cameraman, you are doing it for professional content, not for aesthetics, surprisingly pleasing though it nevertheless turns out to be. During the discussion it is enough just to point the camera at those who speak.

Playing the tape(s) back is easy through the camera on an ordinary television set, perhaps even in your hotel room during the conference. You can make copies for your colleagues through your videomachine from this form of playback. You can play VHS tapes through a videomachine using a VHS adapter. There is unfortunately no adapter for the tiny 8mm Sony tapes, although Sony do make a rather expensive playback machine in this format.

It is at first astonishing how much more one sees, hears and learns through rerunning a talk, especially from easily overlooked comments in the discussion. It is also an admirable way of getting some insight into your own performance if you can persuade a colleague to video your presentation. These tapes are becoming a valuable record of the performances of eminent speakers, and of speakers who may become eminent. How sad it is that we do not have the Oxford debate on Darwin's theories, for example.

Reference

Hawkins CF. Speaking at meetings. In: Hawkins C, Sorgi M. Research: How to plan, speak and write about it. Berlin: Springer-Verlag, 1985; 78–82.

20. Presenting posters at meetings

Let us not speak of them, but look, and pass on!
Dante (1265–1321)

I love posters because, at their best, they are visual, stimulating and quick; a sharp, memorable dose of information (Whimster, 1993). They are not second-class oral presentations; they are not mini-papers; they are a special form of communication for which there are certain criteria. Many posters at many meetings do not meet the criteria, largely because the authors of the posters have had misleading (or no) guidance from organisers who do not know what posters are for. The most reader-unfriendly poster is the one that attempts to reproduce a journal paper – it is not possible to assimilate that amount of information standing in front of a poster.

The message

The hardest part of preparing a poster presentation is deciding what your message is to be; this is not the format for multiple or complicated messages. Many authors do not work out what their message is and just throw everything in, resulting in exasperated viewers. After deciding on the message, the minimum of supporting evidence, preferably pictorial or graphic, is needed for the display. You must be very selective. Thus a reader-friendly poster should convey a clear message, backed up with some evidence, in no more than two minutes, in an attractive way, to people who are on their feet and faced with reading a lot of them. There are five information-transmitting components:

- the title (in the abstract book and heading the display)
- the abstract (in the abstract book)
- the display
- discussion between the author and individual viewers
- poster discussion sessions

The title

The title should indicate the audience(s) at whom the poster is aimed. For example, "Automatised evaluation of nuclear antigens by image analysis" does not make it clear that the poster is about breast cancer, and so may miss the breast cancer people. "Prognostic value of vascularisation in advanced ovarian cancer" is nicely specific about prognosis, vascularisation and cancer of the ovary. The title should be as long as it has to be to include all the target audiences, after attracting the conference organisers, of course.

The abstract

The abstract in the abstract book has not only to persuade the organisers to accept it but also to introduce the display. However, leave something good for denouement in the display.

The abstract in the abstract book should be planned as an integral part of the poster. The participants read the abstract book to decide which posters they want to view and to be introduced to what they are about. The abstract book is a convenient place for retrievable details of methods and results so that these do not have to be copied down by viewers. References too, usually printed in minitype at the bottom of the display, are more conveniently accessible from the abstract.

The display

The display is for the visual part of the presentation with a minimum of print. Such type as there is should be large (many influential, i.e. older, viewers have poor eyesight and wear bifocal spectacles) – can your boss read your poster from a distance of two metres, with the upper, distance, area of his glasses, or must he tip his head up and peer through the lower, close-up, area? (For recommended type sizes see Chapter 7 and Whimster, 1989.) The display should not extend below waist height, because it is undignified to read down there with bifocals. The message should be clear and the supporting evidence clear, brief and easy to follow. Photographs should be large and well labelled and graphs should be large and simple. The aim of conveying the message convincingly within two minutes can be achieved by attention to visual values of colour, design and artistry. "For references see abstract book" can be stated to relieve the viewer from trying to copy them out from tiny print at the bottom of the poster.

The poster is no place for much raw or manipulated data in tables or for much detail about methods, but you can attach a pocket containing handouts or reprints if you must (Chapter 17). Handouts have to be transported to the conference and do not usually do the author any good since they mostly disappear into the bags of those who may never look at them again, while those for whom there are none left feel deprived. You may use various other forms of presentation to enhance the display, for example a video loop to show 3D form better (Whimster and Cookson, 1992), or a novel microscope with a video display visible through the eyepieces (Tucker *et al.*, 1993), or a demonstration of telepathology or telemedicine via a computer link from another centre. Any novel visual enhancement is welcomed by poster viewers, but you have to negotiate with the organisers for suitable space and facilities to mount it.

Discussion with individual viewers

This can be very rewarding. It may be more cost-effective to keep your handouts to hand out personally to viewers who interest you. Some congresses, especially those with so many posters that the space is re-used several times, designate times when poster authors should stand by their posters to talk to viewers. Many authors do not oblige, so it can be difficult to contact them. Make sure you do; you will make new friends and potential collaborators.

Somewhere between discussion with individual viewers and discussion sessions in a lecture theatre is a recent development which takes the form of designated chairmen being available at specified times to take groups of self-selected participants around sets of associated posters (previewed by the chairman) to meet and discuss with the authors. The group of posters may be those in a particular field or speciality, for example echocardiography. This works well if there is room around the poster boards, and if it does not go on so long that people get tired of standing up – a maximum of perhaps 20 minutes – although they can always wander off before that.

Poster discussion session

Many congresses set aside time to discuss posters in a lecture theatre, but the time per poster is usually very short. You cannot expect, and should not be allowed, to give a mini-paper, nor to repeat what has been said in the title, abstract and display. The best and quickest approach is for the chairman to summarise the message, comment briefly on the good and bad points, ask for responses from the audience, and then a rebuttal from the authors. The criterion here is that this is a time for viewers to respond to the authors' efforts. A slide is helpful, but as an aide memoire to the poster, not to insert more information. Chairing a poster session requires the chairman to analyse all the posters for his session, an onerous task that has to be taken seriously.

Prizes

Some conferences award prizes for the best posters. Authors should think what criteria the judges might use when creating their own posters. Obviously they will use the objective referee criteria of "What is the message?"; "Is it original?"; "Is it scientifically sound?' (not very easy to judge from poster material, so substitute "Can the message be believed?"); "Is it of biomedical importance?", to which it is only human to add the subjective criteria "Is it attractive, legible, interesting, imaginative, exciting?".

Practical issues

The poster has to be portable and to be able to fit into and be fixed to the poster panel space designated in advance by the organisers. Check this before you design the poster, and do not make it too big, because encroaching on other people's space causes great offence, or too long, because people do not like reading low down.

There are many ways to make a suitable poster. Some authors make large photographic prints which they carry rolled up in carrying tubes. I use a series of panels small enough to go in a brief case. They were originally made of polystyrene sheet but now consist of laminated A4-size sheets. These can be fixed to anything with double sided tape, which may, however, be difficult or damaging to remove from the panel. Many panels nowadays have a cloth surface to which "Velcro" pads will stick but are easily removed. Conference organisers may provide the materials they want you to use, but this cannot be relied upon. Keep the poster in your hand baggage. Don't forget to collect it afterwards. To adapt Irish advice, "Stick it up early and stick it up often", in whole or in part, at as many meetings as it will bear. The title can be changed and many panels are usable in a variety of settings.

References

Tucker J, Brugal G, Duvall E, Slavin G.HOME – a computerized microscope for use in pathology. Journal of Pathology 1993;170. Supplement:175.

Whimster WF. Wanted: Reader-friendly posters. British Medical Journal 1989;298:274.

Whimster WF, Cookson MJ. Microanatomy of the human lung; confocal microscopy and 3D reconstruction. 7th International Symposium on Diagnostic Quantitative Pathology. Crete, 1992.

Whimster WF. I love posters. Society of Quantitative Pathology Newsletter 1993;14:6–7.

F
THE FINAL WORD

The final word

I suspect that a large part of the formal scientific literature is hardly ever read at all.
John Maddox (*Lancet*, 1968;2:1071)

But it certainly could not be read were it not published, so there's an end on't.
Bill Whimster (1996)

After planning experiments, writing them up as a thesis and as papers for publication in journals, presenting them as posters and oral presentations; after supervising and examining younger players through the same process; and after refereeing, editing, subediting and proofreading other people's papers, hearing their talks and viewing their posters, I am impressed by the infinite variety, subtlety, complexity, value and utterly absorbing nature of the whole up-and-down process. The real heroes are those who have the breakthrough research ideas. I am not one of those, but I have had the good fortune to find myself having to analyse the process, originally for Finnish doctors wishing to publish in English. I have browsed in many books on writing medical and scientific papers and found many useful tips, bits of advice and examples of "good" ways and "bad" ways of doing it, but have ended up dissatisfied with how they might help inexperienced biomedical professionals get through the process competently. I hope this book shows what you can do for yourselves, and makes the essential reference documents readily accessible in the appendices.

You also need to be able to make friends, and stay that way for years, with so many potential helpers: patients, librarians, statisticians, technical staff, computer people, audiovisual people, ethics committee members, grant givers, internal peer reviewers, hospital and university officials, secretaries, journal editors and staff, book publishers, commercial companies, media people and many others. Don't be impatient with them, their preoccupations are different from yours. Don't be brusque or rude, but explain what you are trying to do, thank them, and acknowledge their help in your papers. May your research be blessed with one essential but totally uncontrolled variable, good luck.

Further reading

This section includes useful titles which have been used for reference, but are not quoted in this book.

Albert T. Medical journalism: the writer's guide. Oxford: Radcliffe Medical Press, 1992.

Bryson B. Dictionary of troublesome words. Harmondsworth: Penguin Books, 1984.

Burnard P. Writing for health professionals. London: Chapman and Hall, 1993.

Carey GV. Mind the stop. Harmondsworth: Penguin Books, 1976.

Coggon D, Rose G, Barker DJP. Epidemiology for the uninitiated. 3rd edn. London: British Medical Journal, 1993.

Cox DR, Oakes D. Analysis of survival data. London: Chapman and Hall, 1984.

Friedhof RM. Visualisation: the second computer revolution. New York: Abrams, 1989.

Gore SM, Altman DG. Statistics in practice. London: British Medical Journal, 1982.

Gregory RL, Gombrich EL (eds). Illusion in nature and art. London: Duckworth, 1973.

Hicks W. English for journalists. London: Routledge, 1993.

Johnson FN, Johnson S (eds). Clinical trials. Oxford: Blackwell Scientific, 1977.

Kirkwood BR. Essentials of Medical Statistics. Oxford: Blackwell, 1988.

Matthews DE, Farewell VT. Using and understanding medical statistics. 2nd edn. Basel: Karger, 1988.

Maxwell C. Clinical research for all. Cambridge: Cambridge Medical, 1973.

Miller RG. Survival analysis. New York: John Wiley and Sons, 1981.

Moroney MJ. Facts from figures. 3rd edn. Harmondsworth: Penguin Books, 1956.

Partridge E. Usage and abusage: a guide to good English. London: Book Club Associates, 1978.

Quiller-Couch A. On the art of writing. London: Guild Books Edition, 1946 (1st edn 1916).
Storrie T, Matson J (eds). English usage. London: Cassell, 1994.
Tufte ER. Envisioning information. Connecticut: Graphics Press, 1990.
Whale J. Put it in writing. London: Dent, 1984.
Woodford FP. Scientific writing for graduate students. New York: The Rockefeller University Press, 1968.

APPENDICES

Appendices: Contents

Appendix 1. International Committee of Medical Journal Editors

Preface

In January 1978 a group of editors from some of the major biomedical journals published in English met in Vancouver, British Columbia, and decided on uniform technical requirements for manuscripts to be submitted to their journals. These requirements, including formats for bibliographic references developed for the Vancouver Group by the National Library of Medicine, were published in 1979. The Vancouver group evolved into the International Committee of Medical Journal Editors. Over the years, the group has revised the requirements slightly; this is the fourth edition.

Close to 500 journals have agreed to receive manuscripts prepared in accordance with the requirements. It is important to emphasise what these requirements imply and what they do not.

First, the requirements are instructions to authors on how to prepare manuscripts, not to editors on publication style. (But many journals have drawn on these requirements for elements of their publication styles.)

Second, if authors prepare their manuscripts in the style specified in these requirements, editors of the participating journals will not return manuscripts for changes in style before considering them for publication. Even so, in the publishing process journals may alter accepted manuscripts to conform with details of the journal's publication style.

Third, authors sending manuscripts to a participating journal should not try to prepare them in accordance with the publication style of that journal but should follow the "Uniform Requirements for Manuscripts Submitted to Biomedical Journals".

Authors must also follow the Instructions to Authors in the journal as to what topics are suitable for that journal and the types of papers that may be submitted – for example, original articles, reviews or case reports. In addition, the journal's instructions are likely to contain other requirements unique to that journal, such as number of copies of manuscripts, acceptable languages, length of articles and approved abbreviations.

Participating journals are expected to state in their Instructions to Authors that their requirements are in accordance with the "Uniform Requirements for Manuscripts Submitted to Biomedical Journals" and to cite a published version. This document will be revised at intervals.

A Uniform Requirements for Manuscripts Submitted to Biomedical Journals

Type the manuscript double-spaced, including title page, abstract, text, acknowledgments, references, tables and legends.

Each manuscript component should begin on a new page, in the following sequence: title page, abstract and keywords, text, acknowledgments, references, tables (each table complete with title and footnotes on a separate page) and legends for illustrations.

Illustrations must be good quality, unmounted glossy prints, usually 127 x 173 mm (5 x 7 in.), but no larger than 203 x 251 mm (8 x 10 in.).

Submit the required number of copies of manuscripts and illustrations (see journal's instructions) in a heavy-paper envelope. The submitted manuscript should be accompanied by a covering letter, as described under "Submission of Manuscripts", and permissions to reproduce previously published material or to use illustrations that may identify human subjects.

Follow the journal's instructions for transfer of copyright. Authors should keep copies of everything submitted.

Preparation of manuscript

Type or print out the manuscript on white bond paper, 216 x 279 mm (8.5 x 11 in.), or ISO A4 (210 x 297 mm), with margins of at least 25 mm (1 in.). Type or print on only one side of the paper. Use double-spacing throughout, including title page, abstract, text, acknowledgments, references, individual tables, and legends. Number pages consecutively, beginning with the title page. Put the page number in the upper or lower right-hand corner of each page.

Title Page

The title page should carry: a) the title of the article, which should be concise but informative; b) first name, middle initial and last name of each author, with highest academic degree(s) and institutional affiliation; c) name of department(s) and institution(s) to which the work should be attributed; d) disclaimers, if any; e) name and address of author responsible for correspondence about the manuscript; f) name and address of author to whom requests for reprints should be addressed or statement that reprints will not be available from the author; g) source(s) of support in the form of grants, equipment, drugs, or all these; and h) a short running head or foot line of not more than 40 charactcrs (count letters and spaces) placed at the foot of the title page and identified.

Authorship

All persons designated as authors should qualify for authorship. The order of authorship should be a joint decision of the co-authors. Each author should have participated sufficiently in the work to take public responsibility for the content.

Authorship credit should be based only on substantial contributions to a) conception and design, or analysis and interpretation of data: and to b) drafting the article or revising it critically for important intellectual content: and on c) final approval of the version to be published. Conditions a), b), and c) must all be met. Participation solely in the acquisition of funding or the collection of data does not justify authorship. General supervision of the research group is not sufficient for authorship. Any part of an article critical to its main conclusions must be the responsibility of least one author.

Editors may require authors to justify the assignment of authorship.

Increasingly, multicentre trials are attributed to a corporate author. All members of the group who are named as authors, either in the authorship position below the title or in a footnote, should fully meet the criteria for authorship defined in the "Uniform Requirements". Group members who do not meet these criteria should be listed, with their permission, under Acknowledgements or in an appendix (see "Acknowledgements").

Abstract and Keywords

The second page should carry an abstract (of no more than 150 words for unstructured abstracts or 250 words for structured abstracts). The abstract should state the purposes of the study or investigation, basic procedures (selection of study subjects or laboratory animals; observational and analytical methods), main findings (give specific data and their statistical significance, if possible), and the principal conclusions. Emphasise new and important aspects of the study or observations.

Below the abstract provide, and identify as such, 3 to 10 keywords or short phrases that will assist indexers in cross-indexing the article and that may be published with the abstract. Use terms from the medical subject headings (MeSH) list of *Index Medicus*; if suitable MeSH terms are not yet available for recently introduced terms, present terms may be used.

Text

The text of observational and experimental articles is usually – but not necessarily – divided into sections with the headings Introduction, Methods, Results, and Discussion. Long articles may need subheadings within some sections to clarify their content, especially the Results and Discussion sections. Other types of articles such as case reports, reviews and editorials are likely to need other formats. Authors should consult individual journals for further guidance.

Introduction

State the purpose of the article. Summarise the rationale for the study or observation. Give only strictly pertinent references, and do not review the article extensively. Do not include data or conclusions from the work being reported.

Methods

Describe your selection of the observational or experimental subjects (patients or laboratory animals, including controls) clearly. Identify the methods, apparatus (manufacturer's name and address in parentheses) and procedures in sufficient detail to allow other workers to reproduce the results. Give references to established methods, including statistical methods (see below); provide references and brief descriptions for methods that have been published but are not well known; describe new or substantially modified methods, give reasons for using them, and evaluate their limitations. Identify precisely all drugs and chemicals used, including generic name(s), dose(s) and route(s) of administration.

Ethics

When reporting experiments on human subjects, indicate whether the procedures followed were in accordance with the ethical standards of the responsible committee on human experimentation (institutional or regional) or with thc Helsinki Declaration of 1975, as revised in 1983. Do not use patients' names, initials or hospital numbers, especially in illustrative material. When reporting experiments on animals, indicate whether the institution's or the National Research Council's guide for, or any national law on, the care and use of laboratory animals was followed.

Statistics

Describe statistical methods with enough detail to enable a knowledgeable reader with access to the original data to verify the reported results.When possible, quantify findings and present them with appropriate indicators of measurement error or uncertainty (such as confidence intervals). Avoid sole reliance on statistical hypothesis testing, such as the use of P values, which fails to convey important quantitative information. Discuss eligibility of experimental subjects. Give details about randomisation. Describe the methods for and success of any blinding of observations. Report treatment complications. Give numbers of observations. Report losses to observation (such as dropouts from a clinical trial). References for study design and statistical methods should be to standard works (with pages stated) when possible rather than to papers in which the designs or methods were originally reported. Specify any general-use computer programs used.

Put a general description of methods in the methods section. When data are summarised in the Results section, specify the statistical methods used to analyse them. Restrict tables and figures to those needed to explain the argument of the paper and to assess its support. Use graphs as an alternative to tables with many entries; do not duplicate data in graphs and tables. Avoid nontechnical uses of technical terms in statistics, such as "random" (which implies a randomising device), "normal", "significant", "correlations" and "sample". Define statistical terms, abbreviations. and most symbols.

Results

Present your results in logical sequence in the text, tables and illustrations. Do not repeat in the text all the data in the tables or illustrations; emphasise or summarise only important observations.

Discussion

Emphasise the new and important aspects of the study and the conclusions that follow from them. Do not repeat in detail data or other material given in the Introduction or the Results section. Include in the Discussion section the implications of the findings and their limitations, including implications for future research. Relate the observations to other relevant studies. Link the conclusions with the goals of the study but avoid unqualified statements and conclusions not completely supported by your data. Avoid claiming priority and alluding to work that has not been completed. State new hypotheses when warranted, but clearly label them as such. Recommendations, when appropriate, may be included.

Acknowledgements

At an appropriate place in the article (title page, footnote or appendix to the text: see the journal's requirements), one or more statements should specify a) contributions that need acknowledging but do not justify authorship, such as general support by a departmental chair; b) acknowledgements of technical help; c) acknowledgements of financial and material support, specifying the nature of the support; d) financial relationships that may pose a conflict of interest.

Persons who have contributed intellectually to the paper but whose contributions do not justify authorship may be named and their function or contribution described – for example, "scientific adviser", "critical review of study proposal", "data collection", or "participation in clinical trial". Such persons must have given their permission to be named. Authors are responsible for obtaining written permission from persons acknowledged by name, because readers may infer their endorsement of the data and conclusions.

Technical help should be acknowledged in a paragraph separate from those acknowledging other contributions.

References

Number references consecutively in the order in which they are first mentioned in the text. Identify references in text, tables and legends by Arabic numerals in parentheses. References cited only in tables or in legends to figures should be numbered in accordance with a sequence established by the first identification in the text of the particular table or figure.

Use the style of the examples below, which are based with slight modifications on the formats used by the US National Library of Medicine in *Index Medicus*. The titles of journals should be abbreviated according to the style used in *Index Medicus*. Consult "List of Journals Indexed in *Index Medicus*", published annually as a separate publication by thc library and as a list in the January issue of *Index Medicus*.

Try to avoid using abstracts as references: "unpublished observations" and "personal communications" may not be used as references, although references to written, not oral, communications may be inserted (in parentheses) in the text. Include in the references papers accepted but not yet published; designate the journal and add "In press". Information from manuscripts submitted but not yet accepted should be cited in the text as unpublished observations (in parentheses).

The references must be verified by the author(s) against the original documents.

Examples of correct forms of references are given below.

Articles in journals

1. Standard journal article (list all authors, but if the number exceeds six, give six followed by *et al.*)

> You CH, Lee KY, Chey RY, Menguy R. Electrogastrographic study of patients with unexplained nausea, bloating and vomiting. Gastroenterology 1980 Aug;79(2):311–14.

As an option, if a journal carries continuous pagination throughout a volume, the month and issue number may be omitted.

> You CH, Lee KY, Chey RY, Menguy R. Electrogastrographic study of patients with unexplained nausea, bloating and vomiting. Gastroenterology 1980;79:311–14.

2. Organisation as author

> The Royal Marsden Hospital Bone Marrow Transplantation Team. Failure of syngeneic bone-marrow graft without preconditioning in post-hepatitis marrow aplasia. Lancet 1977;2:742–4.

3. No author given

> Coffee drinking and cancer of the pancreas [editorial]. BMJ 1981;283:628.

4. Article not in English

   Massone L, Borghi S, Pestarino A, Piccini R, Gambini C. Localisations palmaires purpuriques de la dermatite herpetiforme. Ann Dermatol Venereol 1987:114;1545–7.

5. Volume with supplement

   Magni F, Rossoni G, Berti F. BN-52021 protects guinea-pig from heart anaphylaxis. Pharmacol Res Commun 1988;20 Suppl 5:75–8.

6. Issue with supplement

   Gardos G, Cole JO, Haskell D, Marby D, Paine SS, Moore P. The natural history of tardive dyskinesia. J Clin Psychopharmacol 1983:8(4 Suppl):31S–37S.

7. Volume with part

   Hanly C. Metaphysics and innateness: a psychoanalytic perspective. Int J Psychoanal 1988:69(Pt3):389–99.

8. Issue with part

   Edwards L, Meyskens F, Levine N. Effect of oral isotretinoin on dysplastic nevi. J Am Acad Dermatol 1989:20(2 Pt 1):257–60.

9. Issue with no volume

   Baumeister AA. Origins and control of stereotyped movements. Monogr Am Assoc Ment Defic 1978:(3): 353–61.

10. No issue or volume

    Danoek K. Skiing in and through the history of medicine. Nord Medicinhist Arsb 1982:86;100.

11. Pagination in Roman numerals

    Ronne Y, Anvarsfall. Blodtransfusion till fel patient. Vardfacket 1989:13:XXVI–XXVII.

12. Type of article indicated as needed

Spalgo PM, Manners JM. DDAVP and open heart surgery [letter]. Anaesthesia 1989;44:363–4.

Fuhrman SA. Joiner KA. Binding of the third component of complement C3 by Toxoplasma Gondii [abstract]. Clin Res 1987;35:475A.

13. Article containing retraction

Shishido A. Retraction notice: Effect of platinum compounds on murine lymphocyte mitogenesis [Retraction of Alsabti EA, Ghalib ON, Salem MH. In: Jpn J Med Sci Biol 1979;32:53-65]. Jpn J Med Sci Biol 1980:33:235–7.

14. Article retracted

Alsabti EA, Ghalib ON, Salem MH. Effect of platinum compounds on murine lymphocyte mitogenesis [Retracted by Shishido A. In: Jpn J Med Sci Biol 1980:33:235-7]. Jpn J Med Sci Biol 1979;32:53–65).

15. Article containing comment

Piccoli A. Bossatti A. Early steroid therapy in IgA neuropathy: still an open question [comment]. Nephron 1989;51:289–91. Comment on: Nephron 1988;48:12–17.

16. Article commented on

Kobayashi Y, Fujii K, Hiki Y, Tateno S. Kurokawa A, Kamiyama M. Steroid therapy in IgA nephropathy: retrospective study in heavy proteinuric cases [see comments]. Nephron 1988;98:12–17. Comment in: Nephron 1989;51:289–91.

17. Article with published erratum

Schofield A. The CAGE questionnaire and psychological health [published erratum appears in Br J Addict 1989;84:701]. Br J Addict 1988;83:761–4.

Books and Other Monographs

18. Personal author(s)

    Colson JH, Armour WJ. Sports injuries and their treatment. 2nd rev. ed. London: S. Paul. 1986.

19. Editor(s), compiler as author

    Diener HC. Wilkinson M. editors. Drug-induced headache. New York: Springer-Verlag. 1988.

20. Organisation as author and publisher

    Virginia Law Foundation. The medical and legal implications of AIDS. Charlottesville: The Foundation. 1987.

21. Chapters in a book

    Weinstein L. Swartz MN. Pathologic properties of invading microorganisms. In: Sodeman WA Jr. Sodeman WA. editors. Pathologic physiology: mechanisms of disease. Philadelphia: Saunders. 1974:457–72.

22. Conference proceedings

    Vivian VL. editor. Child abuse and neglect: a medical community response. Proceedings of the First AMA National Conference on Child Abuse and Neglect: 1984 Mar 30–31; Chicago. Chicago: American Medical Association. 1985.

23. Conference paper

    Harley NH. Comparing radon daughter dosimetric and risk models. In: Gammage RB. Kaye SB, editors. Indoor air and human health. Proceedings of the Seventh Life Sciences Symposium: 1984 Oct 29–31; Knoxville (TN). Chelsea (MI): Lewis. 1985 :69–78.

24. Scientific or technical report

    Akutsu T. Total heart replacement device. Bethesda (MD): National Institutes of Health. National Heart and Lung Institute; 1971 Apr. Report No.: NIH-NHLI-69-2185-4.

25. Dissertation

Youssef NM. School adjustment of children with congenital heart disease [dissertation]. Pittsburgh (PA): Univ. of Pittsburgh. 1988.

26. Patent

Harred JF, Knight AR, McIntyre JS., inventors. Dow Chemical Company, assignee. Epoxidation process. US patent 3,654.317. 1972 Apr 4.

Other Published Material

27. Newspaper article

Rensberger B. Specter B. CFCs may be destroyed by natural process. The Washington Post 1989 Aug 7; Sect. A2 (col 5).

28. Audiovisual

AIDS epidemic: the physician's role [videorecording]. Cleveland (OH): Academy of Medicine of Cleveland, 1987.

29. Computer file

Renal system [computer program]. MS-DOS version. Edwardsville (KS): MediSim, 1988.

30. Legal material

Toxic Substances Control Act: Hearing on S. 776 before the Subcomm. on the Environment of the Senate Comm. on Commerce. 94th Cong.; 1st Sess. 343 (1975).

31. Map

Scotland [topographic map]. Washington: National Geographic Society (US), 1981.

32. Book of the Bible

Ruth 3:1–18. The Holy Bible. Authorised King James Version. New York: Oxford Univ. Press, 1972.

33. Dictionary and similar references

    Ectasia. Dorland's illustrated medical dictionary. 27th ed. Philadelphia: Saunders, 1988:527.

34. Classical material

    The Winter's Tale: act 5, scene 1, lines 13–16. The complete works of William Shakespeare. London: Rex, 1973.

Unpublished material

35. In press

    Lillywhite HD. Donald JA. Pulmonary blood flow regulation in an aquatic snake. Science. In press.

Tables

Type or print out each table double-spaced on a separate sheet. Do not submit tables as photographs. Number tables consecutively in the order of their first citation in the text and supply a brief title for each. Give each column a short or abbreviated heading. Place explanatory matter in footnotes, not in the heading. Explain in footnotes all non-standard abbreviations that are used in each table. For footnotes use the following symbols, in this sequence: *, †, ‡, §, ‖, ¶, **, ††, ‡‡, ...

Identify statistical measures of variations, such as standard deviation and standard error of the mean.

Do not use internal horizontal and vertical rules.

Be sure that each table is cited in the text.

If you use data from another published or unpublished source, obtain permission and acknowledge fully.

The use of too many tables in relation to the length of the text may produce difficulties in the layout of pages. Examine issues of the journal to which you plan to submit your paper to estimate how many tables can be used per 1,000 words of text.

The editor, on accepting a paper, may recommend that additional tables containing important back-up data too extensive to publish be deposited with an archival service, such as the National Auxiliary Publication Service in the United States, or made available by the authors. In that event an appropriate statement will be added to the text. Submit such tables for consideration with the paper.

Illustrations (figures)

Submit the required number of complete sets of figures. Figures should be professionally drawn and photographed: freehand or typewritten lettering is unacceptable. Instead of original drawings, roentgenograms and other material, send sharp, glossy, black-and-white photographic prints, usually 127 x 173 mm (5 x 7 in.), but no larger than 203 x 254 mm (8 x 10 in.). Letters, numbers and symbols should be clear and even throughout and of sufficient size that, when reduced for publication, each item will still be legible. Titles and detailed explanations belong in the legends for illustrations, not on the illustrations themselves.

Each figure should have a label pasted on its back indicating the number of the figure, author's name, and top of the figure. Do not write on the back of figures or scratch or mar them by using paper clips. Do not bend figures or mount them on cardboard.

Photomicrographs must have internal scale markers. Symbols, arrows or letters used in the photomicrographs should contrast with the background.

If photographs of persons are used, either the subjects must not be identifiable or their pictures must be accompanied by written permission to use the photograph.

Figures should be numbered consecutively according to the order in which they have been first cited in the text. If a figure has been published, acknowledge the original source and submit written permission from the copyright holder to reproduce the material. Permission is required irrespective of authorship or publisher, except for documents in the public domain.

For illustrations in colour, ascertain whether the journal requires colour negatives, positive transparencies or colour prints. Accompanying drawings marked to indicate the region to be reproduced may be useful to the editor. Some journals publish illustrations in colour only if the author pays for the extra cost.

Legends for illustrations

Type or print out legends for illustrations double-spaced, starting on a separate page with Arabic numerals corresponding to the illustrations. When symbols, arrows, numbers or letters are used to identify parts of the illustration, identify and explain each one clearly in the legend. Explain the internal scale and identify the method of staining in photomicrographs.

Units of measurement

Measurements of length, height, weight and volume should be reported in metric units (metre, kilogram or litre) or their decimal multiples. Temperatures should be given in degrees Celsius. Blood pressures should be given in millimetres of mercury. All haematologic and clinical chemistry measurements should be reported in the metric system in terms of the International System of Units (SI). Editors may request that alternative or non-units be added by the authors before publication.

Abbreviations and symbols

Use only standard abbreviations. Avoid abbreviations in the title and abstract. The full term for which an abbreviation stands should precede its first use in the text unless it is a standard unit of measurement.

Submission of manuscripts

Mail the required number of manuscript copies in a heavy-paper envelope, enclosing the manuscript copies and figures in cardboard, if necessary, to prevent bending of photographs during mail handling. Place photographs and transparencies in a separate heavy-paper envelope. Manuscripts must be accompanied by a covering letter signed by all co-authors. This must include:

- information on prior or duplicate publication or submission elsewhere of any part of the work;
- a statement of financial or other relationship that might lead to a conflict of interest;
- a statement that the manuscript has been read and approved by all authors, that the requirements for authorship as previously stated, and furthermore, that each co-author believes that the manuscript represents honest work;
- the name, address and telephone number of the corresponding author, who

is responsible for communicating with the other authors about revisions and final approval of the proofs. The letter should give any additional information that may be helpful to the editor, such as the type of article in the particular journal the manuscript represents and whether the author(s) will be willing to meet the cost of reproducing colour illustrations.

The manuscript must be accompanied by copies of any permissions to reproduce published material, to use illustrations or report sensitive personal information about identifiable persons, or to name persons for their contributions.

Manuscripts on computer disks

For papers that are close to final acceptance, some journals require authors to provide manuscripts in electronic form (on disks) and may accept a variety of word-processing formats or text (ASCII) files.

When submitting disks, authors should:

1. be certain to include a print-out of the manuscript version on the disk;
2. put only the latest version of the manuscript on the disk;
3. name the file clearly;
4. label the disk with the file format and the file name;
5. provide information on hardware and software used.

Authors should consult the journal's Information for Authors for acceptable formats, file- and disk-naming conventions, number of copies to be submitted, and other details.

Participating journals

Journals that have notified the International Committee of Medical Journal Editors of their willingness to consider for publication manuscripts prepared in accordance with earlier versions of the committee's uniform requirements identify themselves as such in their information for authors. A full list is available on request from *Annals of Internal Medicine*.

B Statements of Policy

i. Prior and duplicate publication

Most journals do not wish to consider for publication a paper on work that has already been reported in a published paper or is described in a paper submitted or accepted for publication elsewhere in print or in electronic media. This policy does not usually preclude consideration of a paper that has been rejected by another journal or of a complete report that follows publication of a preliminary report, usually in the form of an abstract. Nor does it prevent consideration of a paper that has been presented at a scientific meeting if not published in full in a proceedings or similar publication. Press reports of the meeting will not usually be considered as breaches of this rule, but such reports should not be amplified by additional data or copies of tables and illustrations. When submitting a paper, an author should always make a full statement to the editor about all submissions and previous reports that might be regarded as prior or duplicate publication of the same or very similar work. Copies of such material should be included with the submitted paper to help the editor decide how to deal with the matter.

Multiple publication – that is, the publication more than once of the same study, irrespective of whether the wording is the same – is rarely justified. Secondary publication in another language is one possible justification, providing the following conditions are met:

1. the editors of both journals concerned are fully informed; the editor concerned with the secondary publication should have a photocopy, reprint or manuscript of the primary version;
2. the priority of the primary publication is respected by a publication interval of at least two weeks;
3. the paper for secondary publication is written for a different group of readers and is not simply a translated version of the primary paper: an abbreviated velsion will often be sufficient;
4. the secondary version reflects faithfully the data and interpretations of the primary version;
5. a footnote on the title page of the secondary version informs readers, peers and documenting agencies that the paper was edited, and is being published, for a national audience in parallel with a primary version based on the same data and interpretations. A suitable footnote might read as follows: "This article is based on a study first reported in the [title of journal, with full reference]."

Multiple publication other than as defined above is not acceptable to editors. If authors violate this rule they may expect appropriate editorial action to be taken.

Preliminary release, usually to public media of scientific information described in a paper that has been accepted but not yet published is a violation of the policies of many journals. In a few cases, and only by arrangement with the editor, preliminary release of data may be acceptable – for example, to warn the public of health hazards.

ii. Retraction of research findings

Editors must assume initially that authors are reporting work based on honest observation. Nevertheless, two types of difficulty may arise.

First, errors may be noted in published articles that require the publication of a correction or erratum of a part of the work. It is conceivable that an error could be so serious as to vitiate the entire body of the work, but this is unlikely and should be handled by editors and authors on an individual basis. Such an error should not be confused with inadequacies exposed by the emergence of new scientific information in the normal course of research. The latter require no corrections or withdrawals.

The second type of difficulty is scientific fraud. If substantial doubts arise about the honesty of a work, either submitted or published, it is the editor's responsibility to ensure that the question is appropriately pursued (including possible consultation with the authors). However, it is not the task of editors to conduct a full investigation or to make a determination; that responsibility lies with the institution where the work has been done or with the funding agency. The editor should be promptly informed of the final decision, and, if a fraudulent paper has been published, the journal must print a retraction.

The retraction, so labeled, should appear in a prominent section of the journal, be listed in the contents page, and include in its heading the title of the original article. It should not simply be a letter to the editor. Ideally, the first author should be the same in the retraction as in the article, although under certain circumstances the editor may accept retractions by other responsible persons. The text of the retraction should explain why the article is being retracted and include bibliographic reference to it.

The validity of previous work by the author of a fraudulent paper cannot be assumed. Editors may ask the author's institution to assure them of the validity of earlier work published in their journals or to retract it. If this is not done they may choose to publish an announcement to the effect that the validity of previously published work is not assured. *(Approved 1987)*

iii. Editorial freedom and integrity

Medical journal owners and editors have a common endeavour, the publishing of a reliable and readable journal, produced with due respect for the stated aims of the journal and for costs. The functions of owners and editors, however, are different. Owners have the right to appoint and dismiss editors and to make important business decisions, in which editors should be involved to the fullest extent possible. Editors must have full authority for determining the editorial content of the journal. This concept of editorial freedom should be resolutely defended by editors even to the extent of placing their positions at stake. To secure this freedom in practice, the editor should have direct access to the highest level of ownership, not only to a delegated manager.

Medical journal editors should have a contract that clearly states the editor's rights and duties in addition to the general terms of the appointment and that defines mechanisms for resolving conflict. An independent editorial advisory board may be useful in helping the editor establish and maintain editorial policy. All editors and editors' organisations have the obligation to support the concept of editorial freedom and to draw major transgressions of such freedom to the attention of the internationial medical community. *(Approved 1988)*

iv. Confidentiality

Manuscripts should be reviewed with due respect for authors' confidentiality. In submitting their manuscripts for review, author's entrust editors with the results of their scientific labour and creative effort, upon which their reputation and career may depend. Authors' rights may be violated by disclosure or by revelation of the confidential details of the review of their manuscript. Reviewers also have rights to confidentiality, which must be respected by the editor. Confidentiality may have to be breached if there are allegations of dishonesty or fraud but otherwise must be honoured.

Editors should not disclose information about manuscripts, including their receipt, their content, their status in the reviewing process, their criticism by reviewers, or their ultimate fate. Such information should be provided only to authors themselves and to reviewers.

Editors should make clear to their reviewers that manuscripts sent for review are privileged communications and are the private property of the authors. Therefore, reviewers and members of the editorial staff should respect the authors' rights by not publicly discussing the authors' work or appropriating their ideas before the manuscript is published. Reviewers should not be allowed to make copies of the manuscript for their files and should be prohibited from sharing it with others except with the permission of the editor. Editors should not keep copies of rejected manuscripts.

Opinions differ on the anonymity of reviewers. Some editors of biomedical journals require their reviewers to sign the comments returned to authors, but most either request that the reviewer's comments not be signed or leave that choice to the reviewer. When comments are not signed the reviewer's identity must not be revealed to the author or anyone else.

Some journals publish reviewers' comments with the manuscript. No such procedure should be adopted without the consent of the authors and reviewers. However, reviewers' comments may be sent to other reviewers of the same manuscript, and reviewers may be notified of the editor's decision. *(Approved 1989)*

v. The role of the correspondence column

All biomedical journals should have a section carrying comments, questions, or criticisms about articles they have published and where the original authors can respond. Usually, but not necessarily, this may take the form of a correspondence column. The lack of such a section denies readers the possibility of responding to articles in the same journal that published the original work. *(Approved 1989)*

vi. Competing manuscripts based on the same study

Editors may receive manuscripts from different authors offering competing interpretations of the same study. They have to decide whether to review competing manuscripts submitted to them more or less simultaneously by different groups or authors, or they may be asked to consider one such manuscript while a competing manuscript has been or will be submitted to another journal. Setting aside the unresolved question of ownership of data, we discuss here what editors ought to do when confronted with the submission of competing manuscripts based on the same study.

Two kinds of multiple submissions are considered: a) submissions by co-workers who disagree on the analysis and interpretation of their study; b) submissions by co-workers who disagree on what the facts are and which data should be reported.

The following general observations may help editors and others dealing with this problem.

Differences in analysis or interpretation

Journals would not normally wish to publish separate articles by contending members of a research team who have differing analyses and interpretations of the data, and submission of such manuscripts should be discouraged. If co-workers cannot resolve their differences in interpretation before submitting a manuscript, they should consider submitting one manuscript containing multiple interpretations and calling their dispute to the attention of the editor so that reviewers can focus on the problem. One of the important functions of peer review is to evaluate the authors' analysis and interpretation and suggest appropriate changes in the conclusions before publication. Alternatively, after the dispute version is published, editors may wish to consider a letter to the editor or a second manuscript from the dissenting authors.

Multiple submissions present editors with a dilemma. Publication of contending manuscripts to air authors' disputes may waste journal space and confuse the reader. On the other hand, if editors knowingly publish a manuscript written by only some of the collaborating team they could be denying the rest of the team their legitirnate co-authorship rights.

Differences in reported methods or results

Workers sometimes differ in their opinions about what was actually done or observed and which data ought to be reported. Peer review cannot be expected to resolve this problem. Editors should decline further consideration of such multiple submissions until the problem is settled. Furthermore, if there are allegations of dishonesty or fraud, editors should inform appropriate authorities.

The cases described above should be distinguished from instances in which independent, non-collaborating authors submit separate manuscripts based on different analyses of publicly available data. In this circumstance editorial consideration of multiple submissions may be justified, and there may even be a good reason for publication of more than one manuscript because different analytical approaches may be complementary and equally valid. *(Approved 1991)*

vii. Order of authorship

The order of authorship is determined by the authors. All authors should meet the basic criteria for authorship (as stated in the "Uniform Requirements"). Because order of authorship is assigned in different ways, its meaning cannot be inferred accurately unless it is stated by the authors. Authors may wish to add an explanation of the order of authorship in a footnote. In deciding on order, authors should be aware that many journals limit the number of authors listed in the table of contents and that the National Library of Medicine lists in MEDLINE only the first 10 authors. *(Approved 1991)*

viii. Guidelines for the protection of patients' right to anonymity

Detailed descriptions or photographs of individual patients, whether of their whole bodies or of body sections (including physiognomies), are sometimes central documentation in medical journal articles. Use of such material may lead to disclosure of a patient's identity, sometimes even indirectly by combination of seemingly innocent information.

Patients (and relatives) have a right to anonymity in published clinical documentation. Details that might identify patients should be avoided unless essential for scientific purposes. Masking of the eye region in photographs of patients may be inadequate protection of anonymity.

If identification of patients is unavoidable, informed consent should be obtained.

Changing data on patients should not be used as a way of securing anonymity.

Medical journals ought to publish their editorial rules for accepting publication of detailed descriptions of individual patients and photographs. When informed consent has been obtained by authors, this should be clearly stated in the article. *(Approved 1991)*

ix. Definition of a peer-reviewed journal

A peer-reviewed journal is one that has submitted most of its published articles for review by experts who are not part of the editorial staff. The numbers and kinds of manuscripts sent for review, the number of reviewers, the reviewing procedures, and the use made of the reviewers' opinions may vary, and therefore each journal should publicly disclose its policies in the Instructions to Authors for the benefit of readers and potential authors. *(Approved 1992)*

x. Medical journals and the popular media

The public's interest in news of medical research has led the popular media to compete vigorously to get information about research as soon as possible. Researchers and institutions sometimes encourage the reporting of research in the popular media before full publication in a scientific journal, by holding a press conference or giving interviews.

The public is entitled to important medical information without unreasonable delay, and editors have a responsibility to do their part in this process. Doctors need to have reports available in full detail, however, before they can advise their patients about the conclusions. In addition, media reports of scientific research before the work has been peer-reviewed and fully published may lead to the dissemination of inaccurate or premature conclusions.

Editors may find the following recommendations useful as they seek to establish policies on these issues.

1. Editors can foster the orderly transmission of medical information from researchers, through peer-reviewed journals, to the public. This can be accomplished by an agreement with authors that they will not publicise their work while their manuscript is under consideration or awaiting publication, and an agreement with the media that they will not release their stories before publication in the journal, in return for which the journal will cooperate with them in preparing accurate stories (see below).
2. Very little medical research has such clear and urgently important clinical implications for the public's health that the news must be released before full publication in a journal. In such exceptional circumstances, however, appropriate authorities responsible for public health should make the decision and should be responsible for the advance dissemination of information to physicians and the media. If the author and the appropriate authorities wish to have a manuscript considered by a particular journal, the editor should be consulted before any public release. If editors accept the need for immediate release, they should waive their policies limiting pre-publication publicity.
3. Policies designed to limit pre-publication publicity should not apply to accounts in the media of presentations at scientific meetings or to the abstracts from these meetings (see "Prior and duplicate publication"). Researchers who present their work at a scientific meeting should feel free

to discuss their presentations with reporters, but they should be discouraged from offering more detail about their study than was presented in their talk.

4. When an article is soon to be published, editors may wish to help the media prepare accurate reports by providing news releases answering questions, supplying advance copies of the journal, or referring reporters to the appropriate experts. This assistance should be contingent upon the cooperation of the media in timing their release of stories to coincide with the publication of the article. *(Approved 1993)*

xi. Conflict of interest

Conflict of interest for a given manuscript exists when a participant in the peer review and publication process – author, reviewer or editor – has ties to activities that could inappropriately influence his or her judgement, whether or not judgement is in fact affected. Financial relationships with industry (for example, employment, consultancies, stock ownership, honoraria, expert testimony), either directly or through immediate family, are usually considered most important conflicts of interest. However, conflicts can occur for other reasons, such as personal relationships, academic competition and intellectual passion.

Public trust in the peer review process and credibility of published articles depend in part upon how well conflict of interest is handled during writing, peer review, and editorial decision-making. Bias can often be identified and eliminated by careful attention to the scientific methods and conclusions of the work. Financial relationships and their effects are less easily detected than other conflicts of interest. Participants in peer review and publication should disclose their conflicting interests, and the information should be made available so that others can judge their effects for themselves. Because readers may be less able to detect bias in review articles and editorials than in reports of original research, some journals do not accept reviews and editorials from authors with a conflict of interest.

Authors

When they submit a manuscript, whether an article or letter, authors are responsible for recognising and disclosing financial and other conflicts of interest that might bias their work. They should acknowledge in the manuscript all financial support for the work and other financial or personal connections to the work.

Reviewers

External peer reviewers should disclose to editors any conflicts of interest that could bias their opinions of the manuscript, and they should disqualify themselves from reviewing specific manuscripts if they believe it appropriate. The editors must be made aware of reviewers' conflicts of interest to interpret the reviews and judge for themselves whether the reviewer should be disqualified. Reviewers should not use knowledge of the work, before its publication, to further their own interests.

Editors and staff

Editors who make final decisions about manuscripts should have no personal financial involvement in any of the issues they might judge. Other members of the editorial staff, if they participate in editorial decisions, should provide editors with a current description of their financial interests, as they might relate to editorial judgements, and disqualify themselves from any decisions where they have a conflict of interest. Published articles and letters should include a description of all financial support and any conflict of interest that, in the editor's judgement, readers should know about.

Editorial staff should not use for private gain the information gained through working with manuscripts. *(Approved 1993)*

Appendix 2. *BMJ*: Advice to authors[1]

(Courtesy of the *BMJ*)

The *BMJ* aims to help doctors everywhere practise better medicine and to influence the debate on health. To achieve these aims we publish original scientific studies, review and educational articles, and papers commenting on the clinical, scientific, social, political, and economic factors affecting health. We are delighted to receive articles for publication in all of these categories from doctors and others. We can publish only about 12% of the articles we receive, but we aim to give quick and authoritative decisions. The editorial staff in London are always happy to advise on submissions by post or telephone.

The *BMJ* is published weekly and has a circulation of about 115,000 of which 20,000 copies are distributed outside Britain. In addition, local editions reach another 150,000 readers. Material published in the weekly journal may be reproduced in these editions and on the Internet (http://www.bmj.com/bmj/).

The *BMJ*'s peer review process

The *BMJ* peer-reviews all the material it receives. About half the original articles are rejected after review in house by two medical editors. The usual reasons for rejection at this stage are insufficient originality, serious scientific flaws, or the absence of a message that is important to a general medical audience. We aim to reach a decision on such papers within two weeks.

---

1 *British Medical Journal* 1996. Points further to the "Uniform Requirements" (Appendix 1A).

The remaining articles are sent to one or more external referees selected from a database of about 2,500 experts. Once returned, those articles thought suitable for publication are discussed by our weekly "hanging committee" of two practising clinicians, two editors, and a statistician.

We aim to reach a final decision on publication within eight weeks of submission. Original articles should be published within three months of being finally accepted after any necessary revisions. We publish six monthly data on how often we achieve these targets.

General instructions

All material submitted for publication must be submitted exclusively to the *BMJ*. All authors must give signed consent to publication. All material should be typed in double line spacing on numbered pages and conform to the "Uniform Requirements for Manuscripts Submitted to Biomedical Journals".

Authors should give their names and their address and appointment at the time they did the work, as well as a current address for correspondence (including telephone and fax numbers). Material accepted for publication will be edited. Proofs are sent to authors of all articles except letters, obituaries, drug points, and contributions to "Medicine and the Media".

Original articles

Please send three copies and keep a further copy for your own reference. We also like accepted articles to be submitted on disk. Our preferred disk format is a file in WordPerfect 5.1 (MS-DOS) on a 3.5-inch disk, but we can also handle versions of WordPerfect and Microsoft Word for DOS, Windows, or Apple Macintosh. Please do not use the software's facilities for automatic page numbering, referencing, or footnotes. Number the pages of the hard copy by hand.

Original papers

Original papers should be no longer than 2,000 words, with a maximum of six tables or illustrations. They should report original research relevant to clinical medicine in a way that is accessible to readers of a general joumal. Papers for the general practice section should cover research or any other matters relevant to primary care. Please include:

- A structured abstract of no more than 250 words with the following headings: objectives, design, setting, subjects, interventions, main outcome measures results, and conclusions
- Original data if you think it will help our reviewers or if we specifically request it
- Copies of related papers you have published. This is particularly important where details of study methods are published elsewhere
- Copies of any non-standard questionnaires used in the research
- Details of sources of funding for the research. These are now published with all papers
- Copies of any previous referees' reports on your research. We appreciate that authors may have tried other journals before sending their work to the *BMJ*. Please let us know how you have responded to previous referees' comments before submission.

Statistical methods should be defined, and any not in common use should be described in detail or supported by references. General guidelines on the use of statistical methods and on the interpretation and presentation of statistical material as well as specific recommendations on statistical estimation and significance have been published.

Papers

Papers may be published with an accompanying commentary, commissioned to help readers interpret research or place it in context. Authors are sent copies of commentaries relating to their work before publication.

Short reports

Short reports must not exceed 600 words, with one table or illustration and five references.

Education and Debate

Education and Debate articles are mostly commissioned, but we welcome reports of up to 2,000 words on all aspects of medicine and health, including sociological aspects of medicine, polemical pieces, and educational articles. These will be peer-reviewed. They should include an unstructured summary of no more than 150 words.

Editorials

Editorials are usually commissioned but we are happy to consider unsolicited editorials of about 80 words. These will be externally peer-reviewed.

Lessons of the week

Lessons of the week are usually case reports or case series alerting readers to a potential clinical problem They should be less than 1,200 words long and accompanied by a single sentence explaining the lesson. The lesson should be as specific as possible and aimed at a general audience.

Other submissions

Letters to the editor

Letters to the editor should be no longer than 400 words with a maximum of five references and one illustration or table. They should be typed in double line spacing and signed by all the authors, who should give their current appointment and address. Priority is given to letters responding to articles published in the joumal within the past four weeks. There is no deadline for letters from abroad, but they should be submitted as soon as possible after publication of the article to which they refer, preferably by fax. All letters from outside Britain are acknowledged. Letters will be edited and may be shortened, but we are unable to send proofs to authors before publication.

Drug points

Drug points usually report new adverse drug reactions or drug interactions. They should be no longer than 300 words with up to five references and one table or figure. Priority will be given to drug points that report more than one case; those in which the patient is rechallenged with the drug; and those which exclude other possible causative factors (disease process, other drugs, environmental agents). We welcome relevant clinical photographs.

Personal View

Personal View articles are always welcome and should be no longer than 1,100 words. We like personal views to be signed, and publish anonymous pieces only by special arrangement.

Medicine and the Media

Contributions to Medicine and the Media are welcome but should be discussed with one of the editors before being submitted.

Obituaries

We will be pleased to receive obituary notices of no more than 80 words, which will be published as soon as possible. We will be pleased to receive, in addition, longer obituaries of up to 400 words. These will be submitted to an editorial committee, which will select some for publication. We welcome good quality photographs and inclusion of the cause of death.

Authorship

As in "Uniform Requirements".

Conflict of interest

We want authors of papers, letters and commissioned articles to let us know whenever they have a conflict of interest capable of influencing their judgements. Such conflicts may take many forms but are likely to be financial, personal, political or academic. Authors should let us know of the potential conflicts even when they are confident that their judgements have not been influenced. We may decide that our readers should know about such a conflict of interest and make up their own minds. Before publishing such information we will consult with the authors. In particular, we want all sources of funding for research to be explicitly stated, and we will include this information in the published paper. We also ask referees to let us know of any conflict of interest.

Patient confidentiality

If there is any chance that a patient might be identified from a case report, illustration or paper, we ask for the written consent of the patient for publication. Patients are almost always willing to give such consent. Black bands across the eyes are wholly ineffective in disguising the patient, and changing details of patients to try to disguise them is bad scientific practice.

Redundant publication

Whenever a paper submitted to the *BMJ* overlaps by more than 10% with previously published papers or papers submitted elsewhere then we want authors to send us copies of those papers. To save readers or researchers from being overwhelmed by redundant material we do not want to publish papers that overlap substantially with papers published elsewhere, and want to make up our own minds on the degree of overlap.

Release of material to the media

We do not want material that is published in the *BMJ* appearing beforehand in other media because doctors and patients are then presented with incomplete material that has not been peer reviewed; they cannot make up their own minds on the validity of the message. We accept that reports may appear in the media after presentations at scientific meetings. Those authors who wish us to publish their papers can clarify matters for the media but should not give them any further information than was included in their presentation. Articles may be withdrawn from publication if given media coverage while under consideration or in press at the *BMJ*.

The *BMJ* issues a press release each week giving details of articles in the forthcoming issue. Authors may be asked to prepare a press release and will be given a chance to approve the final form before release. We write to all authors in the week before publication advising them of possible press interest. Authors approached by journalists in the week before publication should emphasise that all material is embargoed until the Friday morning. The Public Affairs Division of the BMA is happy to advise on the *BMJ* media policy.

Copyright

All authors (with the exception of some government employees) transfer copyright to the *BMJ* just before publication. Readers may make single copies of articles for their own use, but they need permission from the *BMJ* to make multiple copies or to republish substantial parts of the original.

Style

Abbreviations should not be used. Drugs should be referred to by their approved, not proprietary, name and the source of any new or experimental preparations should be given. Scientific measurements should be given in SI units, except for blood pressure which should be expressed in mm Hg.

Tables, illustrations and photographs

Tables and illustrations should be submitted separately from the text of the paper, and legends to illustrations should be typed on a separate sheet. Tables should be simple and should not duplicate information in the text of the paper. Illustrations should be used only when data cannot be expressed clearly in any other way. When graphs, scattergrams or histograms are submitted the numerical data on which they are based should be supplied; in general data given in histograms will be converted into tabular form. Line drawings may be presented as photographic prints or good quality photocopies. Other black and white illustrations should usually be prints, not negatives or x-ray films.

All photographs should be of the highest quality possible as reproduction reduces the quality; they should be no larger than 30 x 21 cm (A4) and should be trimmed to remove all redundant areas; the top should be marked on the back. Colour prints can be used. When photomicrographs are submitted, staining techniques should be stated and an internal scale marker included. If tables or illustrations have been published elsewhere, written consent to re-publication should be obtained by the author from the copyright holder (usually the publisher) and the authors.

X-ray films reproduce most clearly from a glossy black and white print. A slide is unlikely to produce satisfactory results. As well as the photographs you should provide a sketch or tracing paper overlay highlighting the features of interest. Do not mark the photograph. If the feature is small or subtle it may need indicating with an arrow. This should be marked on the tracing paper overlay.

References

This section should be compared with the equivalent section in "Uniform Requirements".

References should be numbered in the order in which they appear in the text. At the end of the article the full list of references should give the names and initials of all authors (unless there are more than six, when only the first six should be given followed by *et al.*). The authors' names are followed by the title of the article; the title of the journal abbreviated according to the style of *Index Medicus* (see "List of Joumals Indexed", printed yearly in the January issue of *Index Medicus*); the year of publication; the volume number; and the first and last page numbers:

> 21 Soter NA, Wassemman Sl, Austen KF. Cold urticaria: release into the circulation of histamine and eosinophil chemotactic factor of anaphylaxis duting cold challenge. *N Eng J Med* 1976;294:687–90.

References to books should give the names of any editors, place of publication, publisher, and year:

> 22 Osler AG. *Complement: mechanisms and functions*. Englewood Cliffs: PrenticeHall, 1976.

Information from manuscripts not yet in press, papers reported at meetings, or personal communications, may be cited only in the text, not as a formal reference. Authors should get permission from the source to cite personal communications. Authors must verify references against the original documents before submitting the article.

Acknowledgement, proofs, press releases and reprints

All original articles will be acknowledged. If the paper is rejected copies will be kept for three months to answer any queries and then shredded. Proof corrections should be kept to a minimum and should conform to the conventions shown in the *Writers' and Artists' Yearbook*. If justifications are needed for corrections to the proofs, please give them in a covering letter, not on the proof. Reprints are available; a scale of charges is included with the proofs.

Guidelines for referees

Referees are asked for their opinion on the originality, scientific reliability, clinical importance, and overall suitability of the paper for publication in the journal, and their reports may be sent to the authors to indicate any changes required. To help them, referees are sent the following guidelines:

- The manuscript is a confidential document. Please do not discuss it even with the author.
- If you want to consult a colleague or junior please discuss this with us first.
- The referee is providing advice to the editors, who (aided by an editorial committee) make the final decision. We will let you know our decision and will normally pass on your anonymised comments to the author.
- Even if we do not accept a paper we would like to pass on constructive comments that might help the author to improve it.
- For this reason please give detailed comments (with references, if appropriate) that will help both the editors to make a decision on the paper and the authors to improve it. Please give your detailed comments on a separate sheet and make your recommendations and any confidential comments to the editor in a covering letter.

The broad aspects that we should like comments on include:

Originality

(truly original or known to you through foreign or specialist publications or through the grape-vine). Originality is our main criterion for papers and case reports.

Scientific reliability

- overall design of study

Patients studied
- adequately described and their condition defined?

Methods
- adequately described?
- appropriate?

Results
- relevant to problem posed?
- credible?
- well presented (including the use of tables and figures)?

Interpretation and conclusions
- warranted by the data?
- reasonable speculation?
- is the message clear?

References
- up to date and relevant?
- any glaring omissions?

Importance (clinical or otherwise) of the work

Suitability for the BMJ *and overall recommendations*

- appropriate for general readership or more appropriate for special journal?
- if not acceptable now can the paper be made so?

Other points

- ethical aspects
- need for statistical assessment
- presentation including writing style

Checklists for statisticians

The statisticians who review *BMJ* papers complete one of two checklists: one is for general papers and the other, which is more detailed, is for papers on clinical trials. These checklists may be sent to the authors.

Checklist for statistical review of general papers

Design features

1. Was the objective of the study sufficiently described?
2. Was an appropriate study design used to achieve the objective?
3. Was there a satisfactory statement given of source of subjects?
4. Was a pre-study calculation of required sample size reported?

Conduct of study

5. Was a satisfactory response rate achieved?

Analysis and presentation

6. Was there a statement adequately describing or referencing all statistical procedures used?
7. Were the statistical analyses used appropriate?
8. Was the presentation of statistical material satisfactory?
9. Were the confidence intervals given for the main results?
10. Was the conclusion drawn from the statistical analysis justified?

Recommendation on paper

11. Is the paper of acceptable statistical standard for publication?
12. If "No" to question 11, could it become acceptablewith suitable revision?

Checklist for statistical review of papers on clinical trials

Design features

1. Was the objective of the trial sufficiently described?
2. Was a satisfactory statement given of diagnostic criteria for entry to the trial?

3. Was there a satisfactory statement given of source of subjects?
4. Were concurrent controls used (as opposed to historical controls)?
5. Were the treatments well defined?
6. Was random allocation to treatment used?
7. Was the method of randomisation described?
8. Was there an acceptably short delay from allocation to start of treatment?
9. Was the potential degree of blindness used?
10. Was there a satisfactory statement of criteria for outcome measures?
11. Were the outcome measures appropriate?
12. Was a pre-study calculation of required sample size reported?
13. Was the duration of post-treatment follow-up stated?

Conduct of trial

14. Were the treatment and control groups comparable in relevant measures?
15. Were a high proportion of the subjects followed up?
16. Did a high proportion of subjects complete treatment?
17. Were the subjects who dropped out from treatment and control groups described adequately?
18. Were side effects of treatment reported?

Analysis and presentation

19. Was there a statement adequately describing or referencing all statistical procedures used?
20. Were the statistical analyses used appropriate?
21. Were prognostic factors adequately considered?
22. Was the presentation of statistical material satisfactory?
23. Were confidence intervals given for the main results?
24. Was the conclusion drawn from the statistical analysis justified?

Recommendation on paper

25. Is the paper of acceptable statistical standard for publication?
26. If "No" to question 25, could it become acceptable with suitable revision?

Drug points

When we receive an item for the drug points page, we usually send the author a checklist of points that should be mentioned.

It is our policy to ask authors who are reporting side effects of drugs to contact the Committee on the Safety of Medicines and the manufacturer of the drug to inquire if they have had similar reports, and to let us have sight of their replies.

Any reports of adverse drug reactions should include the following information:[2]

- Age and sex of subjects.
- Suspected drug and all drugs currently being taken, with start, stop, and restart dates and outcome. (Generally something more than simple coincidence in time is required: rechallenge, with the patient's informed consent, or immunological investigations may tip the balance of probabilities).
- Prior experience with drug or adverse reactions to related drugs.
- Other diseases, environmental factors, and timing.
- Ancillary information from pharmaceutical company and regulatory agency.
- Any published reports.
- Other factors relevant to verify specific types of adverse drug reactions (for example, blood concentration, overdose, baseline laboratory data, ethnic group).
- Please send any pictures we could use to illustrate your drug point.
- Drug points should be under 300 words long with at most five references.

Technical editor's checklist

When returning your revised paper, please supply the information requested below and return the enclosed forms on authorship and conflict of interest.[3]

1. Address for each author and one position held at time of study.
2. Author for correspondence.
3. Structured abstract of 250 words (details enclosed).
4. Opening summary of 75–100 words; lesson of less than 15 words.

---

2 Based on guidelines drawn up at a workshop of representatives of the pharmaceutical industry, departments of clinical pharmacology, drug regulating agencies, medical and scientific editors, and science correspondents of the general press in 1984 (Ciba-Geigy workshop; BMJ 1984;289:898).

3 These forms accompany a paper when it is returned to the author(s) for revision. Also included are details relating to points 14, 15 and 16.

5. Abbreviations should not be used and should be spelt out in full each time.
6. Actual numbers of patients or subjects, as well as percentages, within the text and tables.
7. All values in SI units (except blood pressure in mm Hg).
8. Double spacing (not 1½ spacing) for text and references, margins 3 cm or wider.
9. This article is too long as a short report. Please reduce it to within 600 words with one table or figure and at most five references.
10. Key messages box (details enclosed).
11. Source of funding.
12. References must be set out in Vancouver style (*BMJ* 9 February 1991;302:338–41), and include:
    (a) sumames and initials of all authors (or of only the first six if there are more than six)
    (b) title of the article or chapters
    (c) page numbers of each article or chapter
    (d) editors of books
    (e) publisher of each book
    (f) place of publication of books
    (g) year of publication of books
    (h) title of joumal in full
    (i) volume number of journal
    (j) has the reference been published or accepted for publication? If not please cite in text and renumber other references
13. The numbers from which histograms were drawn. If these are percentages please also provide the actual numbers. (We generally convert histograms into tables, but even if we leave them as histograms the data from which they were drawn are helpful.)
14. Summary of 150 words for "This Week in *BMJ*".
15. Press release.
16. Please supply an electronic version as well as a printed copy.

References

Gardner MJ, Altman DG, eds. Statistics with confidence. London: BMJ, 1989.
Haynes RB, Mulrow CD, Huth EJ, Altman DG, Gardner MJ. More informative abstracts revisited Ann Intern Med 1990;113:69–76.
International Committee of Medical Journal Editors. Protection of patients' rights to privacy. BMJ 1995; 311:1272.
International Committee of Medical Journal Editors. Uniform requirements for manuscripts submitted to biomedical journals. Philadelphia, PA: ICMJE, 1993.
Smith J. Keeping confidences in published papers. BMJ 1992;302:1168.

Appendix 3. Declaration of Helsinki

adopted by the 18th World Medical Assembly, Helsinki 1964
amended by the 29th World Medical Assembly, Tokyo 1975
amended by the 35th World Medical Assembly, Venice 1983

Recommendation for the conduct of clinical research

Introduction

It is the mission of doctors to safeguard the health of the people. His or her knowledge and conscience are dedicated to the fulfilment of this mission.

The Declaration of Geneva of the World Medical Association binds the physician with the words "The health of my patient will be my first consideration" and the International Code of Medical Ethics declares that "A physician shall act only in the patient's interest when providing medical care which might have the effect of weakening the physical and mental condition of the patient".

The purpose of biomedical research involving human subjects must be to improve diagnostic, therapeutic and prophylactic procedures and the understanding of the aetiology and pathogenesis of disease.

In current medical practice most diagnostic, therapeutic and prophylactic procedures involve hazards. This applies especially to biomedical research.

Medical progress is based on research which ultimately must rest in part on experimentation involving human subjects.

In the field of biomedical research a fundamental distinction must be recognised between medical research, in which the aim is essentially diagnostic or therapeutic for a patient, and medical research, the essential object of which is purely scientific, and without implying any direct or therapeutic value to the person subjected to the research.

Special caution must be exercised in the conduct of research which may affect the environment, and the welfare of animals used for research must be respected.

Because it is essential that the results of laboratory experiments be applied to human beings to further scientific knowledge and to help suffering humanity, the World Medical Association has prepared the following recommendations to guide every physician in biomedical research in human subjects. They should be kept under review in the future. It must be stressed that the standards as drafted are only a guide to physicians all over the world. Physicians are not relieved from criminal, civil and ethical responsibilities under the laws of their own countries.

I. Basic principles

1. Biomedical research involving human subjects must conform to generally accepted scientific principles and should be based on adequately performed laboratory and animal experimentation and on a thorough knowledge of the scientific literature.

2. The design and performance of each experimental procedure involving human subjects should be clearly formulated in an experimental protocol which should be transmitted to a specially appointed committee for consideration, comment and guidance.

3. Biomedical research involving human subjects should be conducted only by scientifically qualified persons and under the supervision of a clinically competent medical person. The responsibility for the human subject must always rest with a medically qualified person and never rest on the subject for research, even though the subject has given his or her consent

4. Biomedical research involving human subjects cannot legitimately be carried out unless the importance of the objective is in proportion to the inherent risk to the subject.

5. Every biomedical research project involving human subjects should be preceded by careful assessment of predictable risks in comparison with foreseeable benefits to the subject or to others. Concern for the interests of the subject must always prevail over the interests of science and society.

6. The right of the research subject to safeguard his or her integrity must always be respected. Every precaution should be taken to respect the privacy of the subject and to minimise the impact of the study on the patient's physical and mental integrity and on the personality of the subject.

7. Physicians should abstain from engaging in research projects involving human subjects unless they are satisfied that the hazards involved are

believed to be predictable. Physicians should cease any investigation if the hazards are found to outweigh the potential benefits.

8. In publication of the results of his or her research, the physician is obliged to preserve the accuracy of the results. Reports of experimentation not in accordance with the principles laid down in this Declaration should not be accepted for publication.

9. In any research on human beings, each potential subject must be adequately informed of the aims, methods, anticipated benefits and potential hazards of the study and the discomfort it may entail. He or she should be informed that he or she is at liberty to abstain from participation in the study and that he or she is free to withdraw his or her consent to participation at any time. The physician should then obtain the subject's freely given informed consent, preferably in writing.

10. When obtaining informed consent for the research project the physician should be particularly cautious if the subject is in a dependent relationship to him or her or may consent under duress. In that case the informed consent should be obtained by a physician who is not engaged in the investigation and is completely independent of this official relationship.

11. In the case of legal incompetence, informed consent should be obtained from the legal guardian in accordance with national legislation. Where physical or mental incapacity makes it impossible to obtain informed consent, or when the subject is a minor, permision from the responsible relative replaces that of the subject in accordance with national legislation.
    Whenever the minor child is in fact able to give consent, the minor's consent must be obtained in addition to the consent of the minor's legal guardian.

12. The research protocol should always contain a statement of the ethical considerations involved and should indicate that the principles enunciated in the present Declaration are complied with.

II. Medical research combined with professional care (clinical research)

1. In the treatment of the sick person, the physician must be free to use a new diagnostic and therapeutic measure, if in his or her judgement it offers hope of saving life, re-establishing health or alleviating suffering.

2. The potential benefits, hazards and discomfort of a new method should be weighed against the advantages of the best current diagnostic and therapeutic methods.

3. In any medical study, every patient – including those of a control group, if any – should be assured of the best proven diagnotic or therapeutic method.

4. The refusal of the patient to participate in a study must never interfere with the physician–patient relationship.

5. If the physician considers it essential not to obtain informed consent, the specific reasons for this proposal should be stated in the experimental protocol for transmission to the independent committee.

6. The physician can combine medical research with professional care, the objective being the acquisition of new medical knowledge, only to the extent that medical research is justified by its potential diagnostic or therapeutic value for the patient.

III. Non-therapeutic biomedical research involving human subjects (non-clinical biomedical research)

1. In the purely scientific application of medical research carried out on a human being, it is the duty of the physician to remain the protector of the life and health of that person on whom biomedical research is being carrried out.

2. The subjects should be volunteers – either healthy persons or patients for whom the experimental design is not related to the patients illness.

3. The investigator or the investigating team should discontinue the research if in his, her or their judgement it may, if continued, be harmful to the individual.

4. In research on man, the interest of science and society should never take precedence over considerations related to the wellbeing of the subject.

Appendix 4. Applications for ethical approval and for research grants

The text of edited and condensed application forms for ethical approval and research grants for submission to the Ethics and Research Committees of King's College Hospital and Medical School given here provide biomedical readers realistic examples of the effort they will have to put into getting a project off the ground.

A. Research Ethics Committee protocol proforma

1. INVESTIGATING PERSONNEL
(The Principal Investigator must be a Consultant unless a General Practitioner project; or a Senior Nurse if a Nursing project; or a Senior Officer for other projects.)

PRINCIPAL INVESTIGATOR ....
DEPARTMENT ....

OTHER INVESTIGATORS ....
DEPARTMENT ....

2. TITLE OF PROJECT
Please indicate the Project Title ....
Proposed Start Date ....

3. SUPPORT
Is there any financial support by a drug company or other outside commercial organisation?

If yes, please state which organisation, and does it agree to abide by the ABPI (Association of the British Pharmaceutical Industry) guidelines?

4. OBJECTIVE
(a) What hypothesis is it intended to test?
(b) What is the value of the study to the patient(s) or volunteers?

5. DESIGN OF THE STUDY
Describe briefly, including proposed methods for the analysis of the results.

6. SCIENTIFIC BACKGROUND
(a) If this investigation has been done previously with human subjects, why repeat it?
(b) If it has not been done with humans before, has the problem been worked out as fully as possible in animals, analytically, technically and to assess possible toxic effects?

7. SUBJECTS AND CONTROLS
How will the subjects and controls (if any) be selected and what wider population will they be representative of?
Will pregnancy be excluded?
How many subjects and controls will be involved, and in what age group?
Have sample size calculations been checked with an expert statistician?
Where the analysis involves investigating differences between groups, please give details of the sample size calculations, including the minimum clinically important difference which you wish to be able to detect.

> (e.g. "A sample of 25 patients and 25 controls will be sufficient to detect a difference of 20mm Hg between groups, with a power of 90% and a significance level of 5%")

8. DRUGS
I. If drugs are to be used, then does the drug that is the subject of the investigation have:
(a) A full Clinical Trial Certificate?
(b) A Clinical Trial Exemption Certificate?
(c) If neither (a) nor (b) apply, is the substance being used without a Product Licence for the stated indication?

II. Other Drugs
(a) Please state all other drugs involved in the study:
(b) Are these drugs being supplied by the drug company?

III. Pharmacy Support
(a) Has the Clinical Trials Pharmacist (Ext.3007) been informed?
(b) Where will supplies of drugs be kept?
(c) It is recommended that a copy of the Trial Codes be kept In Parmacy. Do you object, and if so, why?

IV. What efforts will be made to exclude unknown drugs or other unknown medication in patients or volunteers?

V. Substances to be given to subjects
(a) Describe any special diet, isotopic tracers, or other information related to this study.
(b) State routes of administration, amount and effect expected.

9. RADIOACTIVE SUBSTANCES
If radio isotopes are to be used, you are required to register the project with the Radiation Protection Adviser (Department of Medical Physics) and to obtain approval of the Health Ministers through the Department of Health Administration of Radioactive Substances Advisory Committee.
Are radio isotopes to be used in this study?
If radio isotopes are to be used, please indicate that approval has been obtained from the DHSS Administration of Radioactive Substances Advisory Committee
Please provide a copy of the Authority Certificate, and any additional comments received from the Department of Health.

10. SAMPLES TO BE TAKEN FROM THE SUBJECT (Venepuncture, arterial, urine, biopsy etc.)
(a) State type of sample, frequency and amount.
(b) Would the sample(s) be taken especially for this investigation or as part of normal patient care?

11. PROCEDURES
Describe the exact procedures which will be applied to each subject.

12. DISCOMFORT
What discomfort or interference, however slight, with their activities may suffered by all or any of these subjects?

13. SAFETY
Please give details of any potential hazards or side-effects.
Please note that the Ethics Committee should be informed of any adverse reaction or side effects which occur during the course of an approved Trial.

14. INFORMATION TO PATIENTS' GENERAL PRACTITIONERS
(a) Please indicate briefly the information that will be given to GPs about involvement of their patients in the research project. (For drug studies, this should include the name of the active drug, the possible mode of action and known side effects.)
(b) GPs may know of reasons why patients should not participate in the study and a letter should be sent to the GP to ask whether he/she knows of any such reasons.
Please confirm that such a letter will be sent.

15. VOLUNTEERS
(a) Are any payments to be made to volunteers? If so, please give details.
(b) Where it is proposed to recruit medical students or student nurses as volunteers, the supervising authorities must be informed.
If applicable, please confirm that the supervising authorities have been so informed.

16. CONSENT
WRITTEN consent of patients is required in all cases.
Please confirm that such consent will be obtained.

17. PATIENT/PARENT INFORMATION SHEET
(a) A written Information Sheet about the Trial should be given to participants before they are asked to give written consent. The Information Sheet should be submitted to the Ethics Committee with this proforma before approval can be given to the study. Please indicate that this Information Sheet is attached or give reasons why you have not done so.
(b) The patient should be given the opportunity to take away and consider the Information Sheet and sign the consent form later. If this is not practical, the patient should at least be given sufficient time to read and discuss it with relatives if he/she wishes to do so.
Any discussion about the Trial between patient and Investigator should be in person and not by telephone.
(c) Where patients entering a Trial are under 16 years of age, you will be required to obtain the consent of both the patient and the patient's parents or guardian(s). If applicable, please indicate that such consent will be obtained.

18. THE PRINCIPAL INVESTIGATOR IS RESPONSIBLE FOR INFORMING COLLEAGUES AND OTHER GROUPS WHO MAY BE INVOLVED OR AFFECTED BY THE RESEARCH.

ENCLOSURES WHICH MUST ACCOMPANY THE APPLICATION:

I. Completed declaration of funding proforma (see paragraph 3): the original plus one copy are required.
II. Patient Information Sheet (on headed paper, see paragraph 17).
III. Copy of the Consent Form.

The guidelines which accompany this proforma cover:

1. Submission of protocols and supporting papers
2. Approval of Projects by the Committee
3. The Patient Information Sheet
4. Consent Form
5. Indemnity Form
6. Modifications to existing protocols
7. Resubmission of protocols
8. Chairman's action
9. Monitoring of projects
10. Use of drugs
11. Data protection
12. Research in women
13. Research in prisons
14. Research on people with mental disorders
15. Research on children

B. Application to the Research Committee for a new grant

Note: All clinical projects must be approved by the Research Ethics Committee.

1. Applicant's name etc.

2. Short title of project

3. Abstract of research (not more than 250 words)

4. Proposed duration and start date

5. Protocol: this should be no more than 5 pages of A4 paper under the headings:
   - Title of Project
   - Purpose of proposed Investigation
   - Background of Project (including review of relevant literature)

- Plan of Investigation
- Detailed justification for support requested
- References (full title and page numbers)

6. Hospital/Medical School Department in which research will be carried out

7. Is your research being supported by any outside body?
   i. the topic
   ii. the supporting organisation
   iii. the value
   iv. the duration of the project

8. Is this a related project currently being submitted elsewhere?
If so:
   i. to which organisation?
   ii. by what date is a decision expected?

9. Has this application been submitted elsewhere over the past year?
If so:
   i. to which organisation?
   ii. what was the result?

10. Is the proposed research likely to lead to patentable or otherwise exploitable results?
If so, please provide details.

11. If the project has been approved by the Research Ethics Committee, please quote approval number.

12. What statistical advice have you taken?
If none, please state reason.

13. ESTIMATE OF REQUIREMENTS
A. Personnel Requirements
a) It is essential that salary costs are inclusive of all ON-COSTS (i.e. London Weighting, National Insurance, Superannuation). Inaccuracy of figures presented may result in a grant offer being withdrawn. Requests for incremental increases must be stated at the time of application. Claims for incremental rises will not be accepted once the application has been approved.
b) Awards/Salary costs to be quoted as at the time of application. The Comittee will take account of award/salary increases during the financial year in which the grant is funded.

c) Where the name and grade of staff are known, costs at the correct incremental point must be quoted.
d) All awards/salary figures must be verified with the Personnel Division at KCH, the Medical School or the Post-Graduate Administrative Assistant as appropriate.
e) PhD Studentship Awards – please refer to guidelines

B. Special equipment

C. Other expenses

14. CURRICULUM VITAE OF APPLICANT, including:
Recent publications, include papers accepted for publication.
Previous experience in research (with dates)

15. Agreement of Head of Department

16. Agreement to correctness of awards, salaries and gradings by Divisional Personnel Advisor/Finance Officer.

Appendix 5. A selection of typefaces and sizes

8 Point
Serif Typefaces (fonts)

Courier
Some even speak in one language while their slides are written in another; this is also confusing and should be avoided except perhaps for captions on graphs or diagrams.

Times
Some even speak in one language while their slides are written in another; this is also confusing and should be avoided except perhaps for captions on graphs or diagrams.

Palatino
Some even speak in one language while their slides are written in another; this is also confusing and should be avoided except perhaps for captions on graphs or diagrams.

Sans serif Typefaces (fonts)

Helvetica
Some even speak in one language while their slides are written in another; this is also confusing and should be avoided except perhaps for captions on graphs or diagrams.

Univers
Some even speak in one language while their slides are written in another; this is also confusing and should be avoided except perhaps for captions on graphs or diagrams.

10 Point
Serif Typefaces (fonts)

Courier
Some even speak in one language while their slides are written in another; this is also confusing and should be avoided except perhaps for captions on graphs or diagrams.

Times
Some even speak in one language while their slides are written in another; this is also confusing and should be avoided except perhaps for captions on graphs or diagrams.

Palatino
Some even speak in one language while their slides are written in another; this is also confusing and should be avoided except perhaps for captions on graphs or diagrams.

Sans serif Typefaces (fonts)

Helvetica
Some even speak in one language while their slides are written in another; this is also confusing and should be avoided except perhaps for captions on graphs or diagrams.

Univers
Some even speak in one language while their slides are written in another; this is also confusing and should be avoided except perhaps for captions on graphs or diagrams.

12 Point
Serif Typefaces (fonts)

Courier
Some even speak in one language while their slides are written in another; this is also confusing and should be avoided except perhaps for captions on graphs or diagrams.

Times
Some even speak in one language while their slides are written in another; this is also confusing and should be avoided except perhaps for captions on graphs or diagrams.

Palatino
Some even speak in one language while their slides are written in another; this is also confusing and should be avoided except perhaps for captions on graphs or diagrams.

Sans serif Typefaces (fonts)

Helvetica
Some even speak in one language while their slides are written in another; this is also confusing and should be avoided except perhaps for captions on graphs or diagrams.

Univers
Some even speak in one language while their slides are written in another; this is also confusing and should be avoided except perhaps for captions on graphs or diagrams.

14 Point
Serif Typefaces (fonts)

Courier
Some even speak in one language while their slides are written in another; this is also confusing and should be avoided except perhaps for captions on graphs or diagrams.

Times
Some even speak in one language while their slides are written in another; this is also confusing and should be avoided except perhaps for captions on graphs or diagrams.

Palatino
Some even speak in one language while their slides are written in another; this is also confusing and should be avoided except perhaps for captions on graphs or diagrams.

Sans serif Typefaces (fonts)

Helvetica
Some even speak in one language while their slides are written in another; this is also confusing and should be avoided except perhaps for captions on graphs or diagrams.

Type sizes in Times

Good morning. (16 pt)

Good morning. (18 pt)

Good morning. (24 pt)

Good morning. (30 pt)

Good morning. (40 pt)

(all single spacing, interline space set by the programme)

Appendix 6. American and British usage in spelling

The main sources of confusion are ae and oe, both of which are usually kept in British use but almost always contracted to an e in American. Differences that are less important include the American –or for –our, f for ph, and the terminals –ter for –tre, or –er for –re, and –ize for –ise.

The following lists are not intended to be complete; rather, they provide some examples of American–British equivalents that will serve as guides.

	American	*British*
e for ae	etiology	aetiology
	anemia	anaemia
	anesthetic	anaesthetic
	cecum	caecum
	defecation	defaecation
	diarrhea	diarrhoea
	hematuria	haematuria
	pediatric	paediatric
e for oe	celiac	coeliac
	edema	oedema
	esophagus	oesophagus
Suppression of final –al when it follows –ic	physiologic	physiological
Omission of silent endings	program	programme
	catalog	catalogue
	gram	gramme
Terminal –er for –re	center	centre
	fiber	fibre
	liter	litre
f for ph	sulfonamide	sulphonamide
	sulfur	sulphur
k for c	leukocyte	leucocyte
Omission of u when combined with o	color	colour
	tumor	tumour
Miscellaneous	inquire	enquire
	artifact	artefact

Appendix 7. Examples of needless words that cause verbosity

One word instead of several

a decreased number of	fewer
a degree of	some
adjacent to	near
aetiological factor	cause
a large number of	many
a large proportion of	much
a majority of	most
an adequate amount of	enough
an increased amount of	more
anterior aspect	front
a number of	several
at a rapid rate	rapidly
at some future time	later
at the present time	now
a variety of	various
be of the same opinion	agree
be in favour of	support
bring about	cause
by means of	by, with
causal factor	cause
cells of the mononuclear type	mononuclears
circular in outline	circular
come to the conclusion	conclude
come to the same conclusion	agree
conflict of opinion	disagreement
decreased in length	shortened
decreased in thickness	thinned
decreased in weight	lighter
decreased in width	narrower
definitely proved	proved
due to the fact that	because
encountered more frequently	commoner
fatal outcome	death
familiarise oneself	study
fewer in number	fewer
fifty per cent	half
focal areas	foci
for the most part	mainly

give an account of	describe
has been engaged in a study of	has studied
has a tendency to	may
I cannot disguise from myself that it would appear	apparently
in a large number of cases	often
in a lateral direction	sideways
in a paravertebral position	paravertebral
incline to the view	think
in close proximity to	near
in considerable quantities	abundant
increased in length	lengthened
increased in size	enlarged
increased in weight	heavier
increased in width	widened
inferior extremity	leg
in few instances	seldom
in most cases	usually
in some instances	sometimes
in the absence of	without
in the affirmative	yes
in the event that	if
in the neighbourhood of	near
in the not too distant future	soon
in the order of	about
in the vicinity of	near
is capable of	can
is characterised by	shows
is suggestive of	suggests
it is clear that	clearly
it would appear that	apparently
not in accordance with the facts	false
of a mild nature	mild
of common occurrence	common
of considerable size	large
of long standing	old
of the chronic type	chronic
on a previous occasion	before
on no occasion	never
on numerous occasions	often
on one occasion	once
on two occasions	twice
owing to the fact that	because, as
paediatric age group	children
place a major emphasis on	stress
present only in small numbers	scanty
presents a picture similar to	resembles
prior to	before

quite a large quantity of	much
red in colour	red
serves the function of being	is
subsequent to	after
sufficient number of	enough
superior extremity	arm
the predominant number of	most
the vast majority of	most
to all intents and purposes	virtually
try out	test
two equal halves	halves
unanimity of opinion	agreement
undergo transformation	change
was of the opinion that	believed
what is the explanation of	why
with regard to	about

Short phrases for long

an upper intestinal barium study	a barium meal
on an out-patient basis	as an out-patient
adverse climatic conditions	bad weather
the radiographic cardiovascular silhouette	the heart shadow
worthy of trial	worth trying
the majority of authors	most authors
an excessive amount of	too much
an inadequate amount of	too little
an inch in breadth	an inch broad
an inch in length	an inch long
a small number of	a few
assume the erect position	stand up
assume the recumbent position	lie down
at a distance from	away from
a relationship to	related to
cases of short duration	short cases
commonly occurs	is common
created the possibility	made possible
diversity of opinion	many views
have the appearance of	look like
has a course of long duration	is chronic
in almost all instances	nearly always
it has been reported by Jones	Jones reported
it is generally believed	many think
it is possible that the cause is	the cause may be
occasional cases	some cases
of constant occurrence	always present

of sufficient size	large enough
on account of	because of
personally speaking	I think
the general opinion is	many think
the present author believes	I think

Superfluous phrases

Unnecessary qualifying words are another source of verbiage. Authors should, whenever possible, take responsibility rather than hiding behind such terms as:

Authorities agree that...
It is a well-known fact that...
It is recognised that...

Similarly, vague introductory statements may kill whole sentences or paragraphs. The following are typical clichés:

A difference of opinion exists regarding...
Although certainly not a new finding, it is important to point out that...
At the risk of over-simplification...
I have no hesitation in saying...
It has been demonstrated that...
It has been proposed that...
It is interesting to note that...
It is the purpose of this paper...
It should be emphasised that...
One other consideration should be mentioned...
Perhaps, at first sight, it may seem likely that...
There are relatively few studies reported...
Various explanations have been proposed...
We are repeatedly reminded of the necessity for consideration of...

Appendix 8. The distraction removal test kit

Types of distraction

There are four types of distraction:

1. Structural
2. Grammatical
3. Flow and logic
4. Numerical/statistical

Structural distractions

Test kit = 6 questions

1. Why did you start?
2. What did you do?
3. What did you find?
4. What does it mean?
5. What is the message?
6. What is the proof that the message is true?

Grammatical distractions

Test kit = 9 questions

1. What is the subject and verb of each sentence and are they both singular or plural?
2. Is the most important noun the subject?
3. Is the verb in the right tense?
4. If the verb is passive, would the sentence be clearer in the active, and is information being withheld?
5. Is it clear what any pronouns stand for?
6. What noun do any adjectives or adjectival clauses or phrases tell you more about? Could they be left out?
7. What verb do any adverbs or adverbial clauses or phrases tell you more about? Could they be left out?
8. Are the prepositions correctly used with regard to time, place and person?
9. Do any commas help or hinder the sense?

Flow and Logic

Test kit = 5 tests

1. Long sentences and long words
2. Abbreviations
3. Elegant variation
4. Sequencing
6. Paragraphing

Numbers and statistics

Test kit = 5 tests

1. Multiple experiments
2. Text, tables and figures
3. Raw data
4. Showing comparisons
5. Showing that the comparisons are statistically significant and medically or biologically important.

Appendix 9. Proofreading and proofreaders' marks

Typical publishers' instructions for proof checking

Please use RED for typesetter's errors (i.e. departures from the copy-edited typescript) and BLUE or BLACK for all other corrections or alterations, including those that you think are the copy-editor's fault and page cross references. (Use pen, not pencil.)

- Mark all corrections on the set of proofs labelled 'marked set'. (The typesetter or the editorial staff here may have marked queries or alterations on this set.) Please write clearly.
- Keep corrections to an absolute minimum at this stage, since they are expensive (see below) and cause delays. Limit corrections to typesetter's errors and serious mistakes. Make corrections space-for-space if at all possible (i.e. if you need to insert a word see if you can delete a word on the same line to make room for it).
- Lengthy insertions (for example additional or missing references) are best typed on a separate sheet and stapled to the appropriate page of the proof. (Mark clearly on the proof where they should go.)
- Put a circle round anything you write on the proofs that is NOT intended to be typeset (for example, instructions or queries for us or the typesetter).
- Please note that the author has final responsibility for correcting proofs.

Costs

Author: please note! The typesetters will not charge for correcting their own errors (the ones they themselves may have already marked on the proofs and the ones that you may have added in RED pen). However, they will charge for making new corrections. (This is why we ask you to colour code your corrections.) If the bill for the new corrections is more than 10% of the original setting costs, the surplus cost may be passed on to the author.[1] Since corrections are much more expensive to make than original setting, **this would typically allow the author only one free correction line per printed page of the finished book**.

Corrections are expensive to make because the typesetter will need to retypeset the line, paragraph or page again and paste it into position. If corrections are not made space-for-space, several pages may need to be remade. Hand-work of this kind is expensive, and may introduce further errors.

The bill for typesetting (and the division of costs) is done only after all rounds of proof have been corrected and the book is printed. It is at this point that your commissioning editor could approach you about correction charges.

If you feel that some corrections are the result of errors introduced by the copy-editor, please make a note of this on your proofs and we will take this into account.

1 In the case of multi-contributor works, the excess costs cannot usually be passed on in this way; therefore the volume editors and OUP editors reserve the right to disallow corrections from contributors which could incur such costs.

Proof correction marks

If changes have to be made there are standard proof correction marks (fig. A9.1) and some journals will send a copy with the proof. The important point to remember is that although the place where the correction is to go is marked in the line (⁄) the correction itself, the word or letter to be inserted or changed, must be

Mark the text	*In the margin*	*Meaning*
Now (is) is the time	∂	Delete; take out
Now is the ti me	⁀‿	Close up
Now|is the time	#	Insert space
Now ∧ the time	is	Insert word(s)
It is time ∧ We	⊙	Insert period
It is time ∧ but	,/	Insert comma
It is time ∧ we	;	Insert semicolon
The high ∧ energy pump	|–|	Insert hyphen
Smith ∧ 1977 ∧ stated	(/)/	Insert parentheses
Evaluation of ln e ∧	x∨	Insert as superscript
The value of E max	max	Make subscript
The value of	═	Straighten line(s)
all cases. ∧ The value	⌐	Make new paragraph
of most times	(no ¶)	No paragraph-run in
Teh of value is	(tr) //	Transpose
⊏ E_{max}	⊏	Move left as indicated
E_{max} ⊐	⊐	Move right as indicated
(*Now*) is the time	(Rom)	Roman type
now is the time	(cap)	Capital
Smith (1977) said	(s.c.)	Small capitals
Now is The time	(l.c.)	Lower case
Now is the time	(ital)	Italic
now is the time	(cap ital)	Capital italic
Now is the time	(b.f.)	Boldface type
Now is the time	(stet)	Let stand as is

Fig. A9.1. Standard proof correction marks. Note that these may differ slightly elsewhere from the UK.

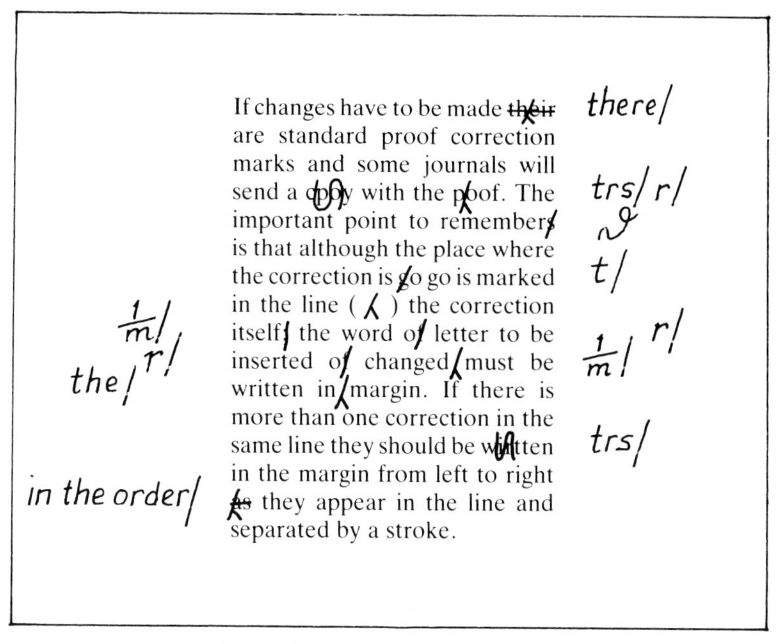

Fig. A9.2. Method of correcting proof.

written in the margin (fig. A9.2). If there is more than one correction in the same line they should be written in the margin from left to right in the order they appear in the line and separated by a stroke. Corrections are put in the margin because the printer will not reread the whole text; he merely looks down the margins and makes corrections only where he sees them. If a word or phrase needs changing, the author should try to replace it with a word or phrase of the same number of characters and spaces, to minimise the amount of resetting that is needed.

If an author notices changes in his wording he should not immediately assume that the editors have got it wrong and change it back but look carefully to see whether the new version is not in fact an improvement. The editors may have picked up ambiguities and rewritten a section to clarify the sense, asking the author to verify it. It is unhelpful merely to reinstate the original wording: that is what caused the problem in the first place, and if the editors in the journal's office are unclear of the meaning the chances are that at least some readers will be too. If the editors have changed the meaning then, of course, the author must correct it and try to write as clearly as he can what he meant. Again, the editors probably got it wrong because it was not completely clear in the first place.

Index